Nathaniel S. Goss

A Revised Catalogue of the Birds of Kansas

With Descriptive Notes of the Nests and Eggs of the Birds Known to Breed

in the State

Nathaniel S. Goss

A Revised Catalogue of the Birds of Kansas
With Descriptive Notes of the Nests and Eggs of the Birds Known to Breed in the State

ISBN/EAN: 9783337140281

Printed in Europe, USA, Canada, Australia, Japan

Cover: Foto ©berggeist007 / pixelio.de

More available books at **www.hansebooks.com**

A REVISED CATALOGUE

OF THE

BIRDS OF KANSAS

WITH DESCRIPTIVE NOTES OF THE NESTS AND EGGS OF THE
BIRDS KNOWN TO BREED IN THE STATE.

By N. S. GOSS.

PUBLISHED UNDER THE DIRECTION OF THE EXECUTIVE COUNCIL,
MAY, 1886.

TOPEKA:
KANSAS PUBLISHING HOUSE: T. D. THACHER, STATE PRINTER.
1886.

PREFACE.

SINCE the publication of my Catalogue of the Birds of Kansas, in 1883, the American Ornithologists' Union have prepared and published a revision of the nomenclature and classification of North American birds, the present accepted, authoritative and standard work. A new edition has therefore become necessary. In the mean time, our knowledge of the birds has increased, and we have become better acquainted with the bird-life of Kansas, especially in the unsettled western portion of the State—a field yet comparatively new to us.

In addition to the description of the nests and eggs of the birds, I have given their times of arrival, and the earliest nests with eggs noticed. The breeding season, however, does not really commence until a little later; and as the State is four hundred miles east and west, and rises from an elevation of about seven hundred and fifty feet to one of about four thousand feet, the times of their arrival must necessarily cover a greater period than would be given to a single locality.

The Catalogue of 1883 embraces 320 species and races; of these, 161 were known to breed in the State. The list, as revised, embraces 335 species and races; of these, 175 are known to breed in the State. A few species and races not appearing in the work have been found both north and south of our limits, and in migration doubtless pass through the State; and it is a surprise to me that the apparently common birds should have so far escaped our notice. The geographical central position of the State makes it a favorable location to catch stragglers and visitants from the adjacent avifaunal provinces, and I feel confident that the list will ultimately reach at least 350.

In this Catalogue, as in the former, I have included only the birds that have come under my own observation, and knowledge gathered from reliable sources. The latter, when new to me, have been duly accredited

in the list.   The descriptions of the nests and eggs are from notes of my
own observations, both in the field and upon the magnificent collection of
eggs made by my brother, Capt. B. F. Goss, and now on display in the
Milwaukee Public Museum, Wisconsin.

The work has been prepared with a view to being of value to the
student interested in the bird-life of our State; and should it meet the
approval of the Executive Council, and of the reader, I shall indeed
feel well repaid for my labor.

TOPEKA, KANSAS, May 1, 1886.

EXPLANATIONS.

1. The five letters, B., R., C., G. and U., each followed by a number, stand respectively for Prof. Spencer F. Baird's Catalogue of 1858; Prof. Robert Ridgway's Catalogue of 1881; Dr. Elliott Coues's Check List of 1882; my Catalogue of 1883; and The American Ornithologists' Union Check List of 1886. The latter has been followed in this list. The dash after any of these letters shows that the bird is not contained in that list.

2. The names on the list, whether of species or subspecies, are consecutively numbered from first to last. The species are distinguished by consisting of two terms, the subspecies of three terms, viz.:

   332. *Merula migratoria.* American Robin.
   333. *Merula migratoria propinqua.* Western Robin.

3. The dimensions of the eggs, as given, represent their average length and diameter in inches and hundredths of an inch.

NOTE.—May 8th, While this work was passing through the press, a female Florida Gallinule was caught near Topeka, and brought to me alive. See No. 67.

BIRDS OF KANSAS.

ORDER PYGOPODES. DIVING BIRDS.

SUBORDER PODICIPEDES. GREBES.

FAMILY PODICIPIDÆ. GREBES.

GENUS COLYMBUS LINNÆUS.

SUBGENUS DYTES KAUP.

B. 706. R. 732. C. 848. G. 317. U. 3.

1. **Colymbus auritus** LINN. Horned Grebe. Migratory; rare. Arrive the middle to last of April.

B. 707. R. 733. C. 850. G. 318. U. 4.

2. **Colymbus nigricollis californicus** (HEERM.). American Eared Grebe. Migratory; rare in eastern, common and may occasionally breed in western Kansas. Arrive the last of April to middle of May.

GENUS PODILYMBUS LESSON.

B. 709. R. 735. C. 852. G. 319. U. 6.

3. **Podilymbus podiceps** (LINN.). Pied-billed Grebe. Summer resident; rare; in migration abundant. Arrive the last of April to first of May. Mr. A. L. Bennett and Mr. V. L. Kellogg, of Emporia, both report finding, May 26th, 1885, in a pond or slough near the city, quite a number of the nests of the birds, containing from five to ten eggs each. Nest in thick weeds or rushes in water from two to three feet deep; composed of old decaying weeds or rushes, brought up from the bottom and piled upon each other in and around the standing stalks, until the fabric reaches the top and floats upon the water, quite a bulky structure. Upon this a small nest is built of debris and bits of slimy moss. Eggs, 1.69x1.17; bluish white at first, but soon become stained in their wet beds; in form elliptical.

SUBORDER CEPPHI. LOONS AND AUKS.

FAMILY **URINATORIDÆ.** LOONS.

GENUS **URINATOR** CUVIER.

B. 698. R. 736. C. 840. G. 320. U. 7.

4. **Urinator imber** (GUNN.). Loon. Migratory; rare. Arrive in April.

ORDER LONGIPENNES. LONG-WINGED SWIMMERS.

FAMILY **LARIDÆ.** GULLS AND TERNS.

SUBFAMILY **LARINÆ.** GULLS.

GENUS **LARUS** LINNÆUS.

B. 661. R. 666a. C. 773. G. 307. U. 51a.

5. **Larus argentatus smithsonianus** COUES. American Herring Gull. Migratory; rare. Arrive in March.

B. 663. R. 668. C. 777. G. 308. U. 53.

6. **Larus californicus** LAWR. California Gull. A rare visitant. One specimen taken by me on the Arkansas river, in Reno county, October 20th, 1880.

B. 664. R. 669. C. 778. G. 309. U. 54.

7. **Larus delawarensis** ORD. Ring-billed Gull. Migratory; quite common. Arrive the last of April to first of May.

B. 668, 669. R. 674. C. 787. G. 310. U. 59.

8. **Larus franklinii** SW. & RICH. Franklin's Gull. Migratory; common. My notes show their capture from the last of March to first of May.

B. 670. R. 675. C. 788. G. 311. U. 60.

9. **Larus philadelphia** (ORD.). Bonaparte's Gull. Migratory; rare. Arrive about the middle of April.

GENUS **XEMA** LEACH.

B. 680. R. 677. C. 790. G. 312. U. 62.

10. **Xema sabinii** (SAB.). Sabine's Gull. A rare accidental visitant. One specimen taken by Mr. Peter Long, at Humboldt, September 21st, 1876.

SUBFAMILY **STERNINÆ.** TERNS.

GENUS **STERNA** LINNÆUH.

SUBGENUS **STERNA.**

B. 686, 691.   R. 685.   C. 798.   G. 313.   U. 69.

11. **Sterna forsteri** NUTT.   Forster's Tern.   Migratory; not uncommon.   Arrive from the middle of April to first of May.   May occasionally breed in the State.

B. 689.   R. 686.   C. 797.   G. 314.   U. 70.

12. **Sterna hirundo** LINN.   Common Tern.   Migratory; very rare.   Arrive from the middle of April to first of May.

SUBGENUS **STERNULA** BOIE.

B. 694.   R. 690.   C. 801.   G. 315.   U. 74.

13. **Sterna antillarum** (LESS.).   Least Tern.   Summer resident; rare; not common in migration.   Arrive the last of April to first of May.   Begin laying about the middle of May.   Nest in a depression or place worked out to fit the body in the sand on the islands and banks of the streams.   Eggs, two to four—rarely ever more than three—1.15x.90; buff to cream white, specked and spotted, in some cases blotched about the large end with brown, umber, and lilac; in form, rather oval to pyriform.

GENUS **HYDROCHELIDON** BOIE.

B. 695.   R. 693.   C. 806.   G. 316.   U. 77.

14. **Hydrochelidon nigra surinamensis** (GMEL.).   Black Tern.   Summer resident; rare; in migration quite common.   Arrive the last of April to first of May. Begin laying the last of May.   Nest on low, wet, marshy ground, bordering ponds and sloughs; made of bits of stems of reeds and grasses, and lined with the leaves and finer stems.   In some cases the eggs are laid upon the bare ground. Eggs, usually three—occasionally four—1.30 x .96, greenish drab to olive brown' spotted and blotched with brownish black, often thickest and running together around large end ; in form rather oval to pyriform.

ORDER STEGANOPODES. TOTIPALMATE SWIMMERS.

FAMILY **ANHINGIDÆ.** DARTERS.

GENUS **ANHINGA** BRISSON.

B. 628.   R. 649.   C. 760.   G. 306.   U. 118.

15. **Anhinga anhinga** (LINN.).   Anhinga.   A rare summer visitant.   In August, 1881, a specimen was captured in the Solomon Valley by C. W. Smith, Esq., of Stockton; and identified by Prof. F. H. Snow, who has the skin of the bird in his cabinet.

FAMILY PHALACROCORACIDÆ. CORMORANTS.

GENUS PHALACROCORAX BRISSON.

SUBGENUS PHALACROCORAX.

B. 623.   R. 643.   C. 751.   G. 304.   U. 120.

16. **Phalacrocorax dilophus** (Sw. & RICH.). Double-crested Cormorant. Migratory; not uncommon. Arrive the last of March to first of April. To be looked for in the old deep channels of the rivers in the low-timbered lands.

B. 625.   R. 644.   C. 754.   G. 305.   U. 121.

17. **Phalacrocorax mexicanus** (BRANDT). Mexican Cormorant. Prof. Snow in his catalogue of the birds of Kansas, says: "Migratory; rare; a single specimen taken four miles south of Lawrence April 2, 1872, by George D. Allen."

FAMILY PELECANIDÆ. PELICANS.

GENUS PELECANUS LINNÆUS.

SUBGENUS CYRTOPELICANUS REICHENBACH.

B. 615.   R. 640.   C. 748.   G. 303.   U. 125.

18. **Pelecanus erythrorhynchos** GMEL. American White Pelican. Migratory; quite common. Arrive the last of April to first of May.

FAMILY FREGATIDÆ. MAN-O'-WAR BIRDS.

GENUS FREGATA CUVIER.

B. 619.   R. 639.   C. 761.   G —.   U. 128.

19. **Fregata aquila** (LINN.). Man-o'-War Bird. A straggler. Mr. Frank Lewis, of Downs, Kansas, reports to me the capture of the bird on the north fork of the Solomon river, in Osborne county, August 16th, 1880. It was killed with a stone, while sitting on a tree. The specimen has passed out of his hands; but he sends me a photograph of the bird, taken after it was mounted, which removes all previous doubts as to its identification. The birds are strictly maritime, and largely parasitical in habits. Their home is on the coast of tropical and subtropical America. They are known to be great wanderers along the sea-board; but this is, I think, the first record of its being found away from the coast-range, and to straggle so far inland it must surely have been crazed or bewildered.

ORDER ANSERES. LAMELLIROSTRAL SWIMMERS.

FAMILY ANATIDÆ. DUCKS, GEESE, AND SWANS.

SUBFAMILY MERGINÆ. MERGANSERS.

GENUS MERGANSER BRISSON.

B. 611. R. 636. C. 743. G. 300. U. 129.

20. Merganser americanus (CASS.). American Merganser. Winter sojourner; quite common. Leave the last of February to middle of March.

B. 612. R. 637. C. 744. G. 301. U. 130.

21. Merganser serrato (LINN.). Red-breasted Merganser. Winter visitant; rare.

GENUS LOPHODYTES REICHENBACH.

B. 613. R. 638. C. 745. G. 302. U. 131.

22. Lophodytes cucullatus (LINN.). Hooded Merganser. Resident; rare; common in winter. Begin laying the last of April. Nest in shallow holes and trough-like cavities in trees near the streams, said to be lined with grass, leaves and down. Eggs, six to ten; 2.10x1.72; pure ivory white, with a few neutral tints; in form rounded oval.

SUBFAMILY ANATINÆ. RIVER DUCKS.

GENUS ANAS LINNÆUS.

B. 576. R. 601. C. 707. G. 282. U. 132.

23. Anas boschas LINN. Mallard. Resident; rare; in migration abundant. Begin laying the first of May. Nest on the ground in the grass, at the edge of ponds or sloughs, constructed of grass, weeds and leaves loosely scraped or placed together, and lined with feathers and down. Eggs, six to ten; 2.30x1.60; dull greenish white, in form elliptical.

B. 577. R. 602. C. 708. G. 283. U. 133.

24. Anas obscura GMEL. Black Duck. Entered in first catalogue as "migratory; rare;" but since, on comparing the specimens captured in the State *that I have seen* with Eastern ones, they prove to be the "Florida Duck." Other writers claim that the birds have been taken in the State, also in Texas, and west to Utah, and I am inclined to think that further investigation will prove it to be the case. With this explanation I let the bird stand as first entered.

B. —. R. 603. C. 709. G. —. U. 134.

25. Anas fulvigula RIDGW. Florida Duck. Migratory; rare. Arrive about the middle of March. I captured a female at Neosho Falls March 11, 1876, and have since shot one, and observed two others in the State. The birds were entered in first catalogue as *A. Obscura.*

SUBGENUS **CHAULELASMUS** BONAPARTE.

B. 584. R. 604. C. 711. G. 284. U. 135.

26. Anas strepera LINN. Gadwell. Summer resident; rare; in migration common. Arrive the middle of March to first of April. Begin laying the last of May. Nesting habits the same as the Mallard. Eggs, a little smaller and paler.

SUBGENUS **MARECA** STEPHENS.

B. 585. R. 607. C. 713. G. 286. U. 137.

27. Anas americana GMEL. Baldpate. Summer resident; very rare; in migration common. Arrive about the middle of March. I have never been so fortunate as to find their nest, or see their eggs. They are said to build under a bush, or bunch of grass, on high ground, often quite a distance from the water — a depression in the ground, lined with leaves and down. Eggs, eight to twelve; average dimensions, 2.10x1.50; creamy white. In form elliptical.

SUBGENUS **NETTION** KAUP.

B. 579. R. 612. C. 715. G. 290. U. 139.

28. Anas carolinensis GMELIN. Green-winged Teal. Winter sojourner; rare; in migration abundant. Leave in April.

SUBGENUS **QUERQUEDULA** STEPHENS.

B. 581. R. 609. C. 716. G. 288. U. 140.

29. Anas discors LINN. Blue-winged Teal. Summer resident; rare; in migration abundant. Arrive the last of March to middle of April. Begin laying the last of May. Nest on the ground in coarse grass, reeds or rushes bordering the prairie sloughs; composed of the same material, and lined with down. Eggs, eight to twelve; 1.86x1.32; in form elliptical to oval.

B. 582. R. 610. C. 717. G. 289. U. 141.

30. Anas cyanoptera VIEILL. Cinnamon Teal. Migratory; very rare in eastern, but not uncommon in middle and western Kansas. Arrive the middle to last of April. June 3d, 1885, I found a pair on a pond in Meade county. From their actions was led to think they had a nest near by; failing to find it, I shot the female, and on dissection found several well-developed eggs in the ovary. I cannot, upon this, safely enter the bird as a summer resident, but I am strong in the belief that they do occasionally breed in the western part of the State.

GENUS **SPATULA** BOIE.

B. 583. R. 608. C. 718. G. 287. U. 142.

31. Spatula clypeata (LINN.). Shoveller. Summer resident; rare; in migration common. Arrive the middle of March to first of April. Begin laying the last of May. Nest near the water, on the ground in a depression or place worked out to fit the body, made of grass and lined with down. Eggs, six to ten; 2.12x1.50; greenish white to pale drab; in form elliptical.

GENUS DAFILA STEPHENS.

B. 578. R. 605. C. 710. G. 285. U. 143.

32. Dafila acuta (LINN.). Pintail. Migratory; common. Arrive the last of February to first of March.

GENUS AIX BOIE.

B. 587. R. 613. C. 719. G. 291. U. 144.

33. Aix sponsa (LINN.). Wood Duck. Summer resident; common. Arrive the last of March to first of April. Begin laying the last of April. Nest in holes in trees on or near the banks of streams, usually in a trough-like cavity of a large broken limb, lined sparingly with grass, weeds or leaves, and a few feathers with down. Eggs, six to fifteen; 2.00x1.50; cream to buff white, smoothly polished; in form, elliptical to oval.

GENUS AYTHYA BOIE.

B. 591. R. 618. C. 723. G. 296. U. 146.

34. Aythya americana (EYT.). Redhead. Migratory; common. Arrive the first of March to middle of April.

B. 592. R. 617. C. 724. G. 295. U. 147.

35. Aythya vallisneria (WILS.). Canvas-back. Migratory; irregular; not uncommon. Arrive early in March. My notes show the capture of one February 22d.

SUBGENUS FULIGULA STEPHENS.

B. 588. R. 614. C. 720. G. 292. U. 148.

36. Aythya marila nearctica STEJN. American Scaup Duck. Migratory; rare. Arrive in March to first of April.

B. 589. R. 615. C. 721. G. 293. U. 149.

37. Aythya affinis (EYT.). Lesser Scaup Duck. Migratory; quite common. Arrive early in March to first of April.

B. 590. R. 616. C. 722. G. 294. U. 150.

38. Aythya collaris (DONOV.). Ring-necked Duck. Migratory; common. Arrive very early. My notes show their capture from February 9th to May 24th.

GENUS GLAUCIONETTA STEJNEGER.

B. 593. R. 620. C. 725. G. 297. U. 151.

39. Glaucionetta clangula americana (BONAP.). American Golden-eye. Mirgratory; rare. Begin to arrive about the middle of March; have seen them as late as the last of April.

GENUS CHARITONETTA STEJNEGER.

B. 595. R. 621. C. 727. G. 298. U. 153.

40. Charitonetta albeola (LINN.). Buffle-head. Migratory; irregular; at times quite common. Arrive the last of February to first of April.

GENUS **ERISMATURA** BONAPARTE.

B. 609.   R. 634.   C. 741.   G. 299.   U. 167.

41. Erismatura rubida (WILS.). Ruddy Duck. Migratory; quite common. Arrive the last of March to last of April.

SUBFAMILY **ANSERINÆ.** GEESE.

GENUS **CHEN** BOIE.

B. 564.   R. 590.   C. 694.   G. 276.   U. —.

[**Chen cærulescens** (LINN.). Blue Goose. This bird has been dropped from the body of the A. O. U. list, and included in the hypothetical list, on account of the possibility, if not probability, that it is a colored phase of *Chen hyperborea.* The plumage of the specimens that have come under my observation in both the adult and young stages, is certainly very distinct from *C. hyperborea*, and in my opinion the bird will eventually be declared a valid species, and restored to the list.]

B. —.   R. 591a.   C. 696.   G. 277.   U. 169.

42. Chen hyperborea (PALL.). Lesser Snow Goose. Abundant in migration. A few occasionally linger into winter. Arrive early in March.

GENUS **ANSER** BRISSON.

B. 565, 566.   R. 593a.   C. 693.   G. 278.   U. 171a.

43. Anser albifrons gambeli (HARTL.). American White-fronted Goose. Migratory; common. Arrive in March.

GENUS **BRANTA** SCOPOLI.

B. 567.   R. 594.   C. 702.   G. 279.   U. 172.

44. Branta canadensis (LINN.). Canada Goose. Common migrants. A few remain during the winter, retiring only when the extreme cold weather closes their watery resorts. Leave in March.

B. 569.   R. 594a.   C. 704.   G. 280.   U. 172a.

45. Branta canadensis hutchinsii (SW. & RICH.). Hutchins' Goose. Migratory; abundant. A few linger into winter. Leave in March.

B. 570.   R. 595.   C. 700.   G. 281.   U. 173.

46. Branta bernicla (LINN.). Brant. Rare; accidental migrants.

SUBFAMILY **CYGNINÆ.** SWANS.

GENUS **OLOR** WAGLER.

B. 561a.   R. 588.   C. 689.   G. 274.   U. 180.

47. Olor columbianus (ORD.). Whistling Swan. Migratory; rare. Arrive about the middle of March.

B. 562.   R. 589.   C. 688.   G. 275.   U. 181.

48. Olor buccinator (RICH.). Trumpeter Swan. Migratory; rare. Arrive about the middle of March.

ORDER HERODIONES. HERONS, STORKS, IBISES, ETC.

SUBORDER IBIDES. SPOONBILLS AND IBISES.

FAMILY **IBIDIDÆ.** IBISES.

GENUS PLEGADIS KAUP.

B. 500a.   R. 504.   C. 650.   G. 234.   U. 187.

49. **Plegadis guarauna** (LINN.). White-faced Glossy Ibis. A rare visitant. Shot at a lake near Lawrence by W. L. Bullene, in the fall of 1879, and reported to me by Prof. F. H. Snow, who has the specimen in the State University.

SUBORDER CICONIÆ. STORKS, ETC.

FAMILY **CICONIIDÆ.** STORKS AND WOOD IBISES.

SUBFAMILY TANTALINÆ. WOOD IBISES.

GENUS TANTALUS LINNÆUS.

B. 497.   R. 500.   C. 648.   G. 233.   U. 188.

50. **Tantalus loculator** LINN. Wood Ibis. Irregular summer visitant; rare. Dr. George Lisle wrote me in the spring of 1883 that he had noticed the birds a few times on the flats east of Chetopa, and that Albert Garrett killed a very fine specimen there about six years ago; and Dr. Lewis Watson, of Ellis, informs me that one put in an appearance March 26th, 1885, and stayed about his premises on the creek for several days. An accidental straggler; a rare find so far north, and so early.

SUBORDER HERODII. HERONS, EGRETS, BITTERNS, ETC.

FAMILY **ARDEIDÆ.** HERONS, BITTERNS, ETC.

SUBFAMILY BOTAURINÆ. BITTERNS.

GENUS BOTAURUS HERMANN.

SUBGENUS BOTAURUS.

B. 492.   R. 497.   C. 666.   G. 231.   U. 190.

51. **Botaurus lentiginosus** (MONTAG.). American Bittern. Summer resident; common. Arrive the last of April to first of May. Begin laying about the

20th of May. Nest on the ground in low marshy places, built upon hummocks in the thickly-growing water grasses, or upon the tops of old broken-down rushes, quite bulky, composed of small sticks, weeds and grasses, or of rushes bitten off about fifteen inches in length and loosely woven together. Eggs, said to be three to six, (I have never found over four in a nest); 2.00x1.48; brownish drab, one set olive drab; in form oval to elliptical.

SUBGENUS **ARDETTA** GRAY.

B. 491. R. 498. C. 667. G. 232. U. 191.

52. Botaurus exilis (GMEL.). Least Bittern. Summer resident; rare; in migration common. Arrive about the first of May. Begin laying the last of May. Nest in rushes and a coarse, tall cane like water grass — a platform about eighteen inches from the ground, or water, made of the stems and leaves woven in and around the standing growing stalks. Eggs, usually four; 1.25x.98; with a faint greenish-blue tinge; in form rounded oval.

SUBFAMILY **ARDEINÆ.** HERONS AND EGRETS.

GENUS **ARDEA** LINN.

SUBGENUS **ARDEA.**

B. 487. R. 487. C. 655. G. 224. U. 194.

53. Ardea herodias LINN. Great Blue Heron. Summer resident; rare; in migration common. Arrive early in March. Begin laying the last of March. Nest in high trees, along the streams; in localities destitute of trees, upon bushes and upon the ground. A flat, bulky structure of sticks, lined sparingly with grasses. Eggs, 2.63x1.82; pale, greenish blue; rather elliptical in form.

SUBGENUS **HERODIAS** BOIE.

B. 486, 486a. R. 489. C. 658. G. 225. U. 196.

54. Ardea egretta GMEL. American Egret. Summer visitant; not uncommon. Arrive from the south in July and August, return in September.

SUBGENUS **GARZETTA** KAUP.

B. 485. R. 490. C. 659. G. 226. U. 197.

55. Ardea candidissima GMEL. Snowy Heron. Summer visitant; not uncommon. Arrive from the south in July and August, return in September.

SUBGENUS **FLORIDA** BAIRD.

B. 490. R. 493. C. 662. G. 227. U. 200.

56. Ardea cœrulea LINN. Little Blue Heron. Summer visitant; rare. Arrive from the south in July and August, return in September.

SUBGENUS **BUTORIDES** BLYTH.

B. 493. R. 494. C. 663. G. 228. U. 201.

57. Ardea virescens LINN. Green Heron. Summer resident; abundant. Arrive about the middle of April. Begin laying about the first of May. Nest placed

on the branches of trees or upon bushes along the streams, made loosely of sticks
and lined with twigs. Eggs, four or five; 1.55x1.15; light greenish blue; in form ·
elliptical to oval.

GENUS **NYCTICORAX** STEPHENS.

SUBGENUS **NYCTICORAX**.

B. 495. R. 495. C. 664. G. 229. U. 202.

58. Nycticorax nycticorax nævius (BODD.). Black-crowned Night Heron.
Summer resident; rare. Arrive about the middle of April. Begin laying about
the middle of May. Nest on the branches of trees, or, in places destitute of trees,
upon bushes and on the ground in thick growths of weeds and small bushes.
Composed of sticks loosely woven together; when built upon the ground, of
sticks, weeds, and grasses. Eggs, three to five; 2.00x1.50; pale greenish blue;
in form elliptical to oval.

SUBGENUS **NYCTHERODIUS** REICHENBACH.

B. 496. R. 496. C. 665. G. 230. U. 203.

59. Nycticorax violaceus (LINN.). Yellow-crowned Night Heron. Summer
resident; rare. Arrive about the middle of April. Begin laying about the middle of May. Nest on trees and bushes. Composed of sticks loosely woven
together. Eggs, three to five; 1.95x1.45; pale yellowish to greenish blue; in
form oval.

ORDER PALUDICOLÆ. CRANES, RAILS, ETC.

SUBORDER **GRUES**. CRANES.

FAMILY **GRUIDÆ.** CRANES.

GENUS **GRUS** PALLAS.

B. 478. R. 582. C. 668. G. 272. U. 204.

60. Grus americana (LINN.). Whooping Crane. Migratory; rare. Arrive about
the middle of March to first of April.

B. 479. R. 583. C. 669. G. 273. U. 206.

61. Grus mexicana (MÜLL.). Sandhill Crane. Migratory; common. Arrive about
the middle of March to first of April.

SUBORDER RALLI. RAILS, GALLINULES, COOTS, ETC.

FAMILY **RALLIDÆ.** RAILS, GALLINULES, AND COOTS.

SUBFAMILY **RALLINÆ.** RAILS.

GENUS **RALLUS** LINNÆUS.

B. 542. R. 569. C. 676. G. 267. U. 208.

62. Rallus elegans AUD. King Rail. Summer resident; common in eastern Kansas. Arrive the first to middle of April. Begin laying about the middle of May. Nest on the ground in marshy places at or near the edge of water, generally upon a hummock in a thick, heavy growth of grass, or under a bush, made of coarse grasses, weeds and rushes, quite bulky, and so woven together as to often form a partial cover overhead. Eggs, six to twelve; 1.63x1.25; ·pale bluish to cream white, sparingly specked and spotted with various shades of reddish brown, and shell stains of purple and lilac, the spots thickest and often running together around large end; in form oval.

B. 554. R. 572. C. 677. G. 268. U. 212.

63. Rallus virginianus LINN. Virginia Rail. Summer resident; rare; during migration common. Arrive the middle of April to first of May. Begin laying about the middle of May. Nest in thick growth of grass on low boggy grounds, quite bulky, made of grass, weeds, etc. Eggs, six to ten; they are said to average 1.25x.95; measurement of a set collected May 21st, 1878, at Pewaukee, Wisconsin, 1.26x.90, 1.27x.90, 1.27x.90, 1.32x.90, 1.28x.91, 1.30x.92, 1.32x.92, 1.35x.92, 1.30x.93, 1.29x.95; cream white, thinly spotted with reddish brown, and faint markings of lilac; thickest around large end; in form oval.

GENUS **PORZANA** VIEILLOT.

SUBGENUS **PORZANA.**

B. 555. R. 574. C. 679. G. 269. U. 214.

64. Porzana carolina (LINN.). Sora. Summer resident; rare; in migration abundant. Arrive the middle of ·April to first of May. Begin laying about the middle of May. Nest on marshy ground, at the border of ponds, and old channels of streams, in elevated tussocks of grass, a shallow or platform nest made loosely of grass, weeds and rushes. Eggs, six to ten; 1.20x.90; grayish to olive drab, specked and spotted with purple and reddish brown; in form oblong oval.

SUBGENUS **COTURNICOPS** BONAPARTE.

B. 557. R. 575. C. 680. G. —. U. 215.

65. Porzana noveboracensis (GMEL.). Yellow Rail. Summer resident; rare. Prof. L. L. Dyche, Curator of Birds and Mammals, State University, writes me that April 18, 1885, he captured one of the birds, (a female,) on low wet land about five miles southeast of Lawrence. The specimen is mounted in the fine collection under his charge. It is the first bird to my knowledge captured, or seen in the State; but this is not strange, as the birds inhabit the marshy grounds, and at the least alarm, run, skulk and hide in the reeds or grass, and it is next to impossible to force them to take wing. Therefore seldom seen even where known to be common. I enter the bird as summer resident, because they have been found both north and south of us, and are known to breed within their geographical range. Nest on the ground. The following description is from Vol. 1, North American Water Birds: "Its nest resembles the ordinary loosely constructed one of this family." * * * "Three eggs in the Smithsonian Collection (No. 7057), from Winnebago, in northern Illinois, measure respectively 1.08 inches by .85, 1.12 by .82, 1.12 by .80. They are of oval shape, one end slightly more tapering than the other. Their ground-color is a very deep buff, and one set of markings, which are almost entirely confined to the larger end, consists of blotches of pale diluted purplish brown; these are overlain by a dense sprinkling of fine dottings of rusty brown." NOTE.—October 1st, Professor Dyche captured on the Wakarusa bottom lands, two and a half miles south of Lawrence, another of the little birds, a female, and he thinks a *young* bird. The lucky finds were both caught by his dog.

SUBGENUS **CRECISCUS** CABANIS.

B. 556.   R. 576.   C. 681.   G. 270.   U. 216.

66. Porzana jamaicensis (GMEL.).   Black Rail.   Summer resident; rare. Arrive about the first of April. Begin laying about the middle of May. Nest in a depression on marshy ground, composed of grass blades; in form, round and deep. Eggs, six to ten; 1.02x.80; creamy white, thickly sprinkled with small dots of reddish brown; in form, oval. Two eggs—the remains of a set of eight collected near Manhattan, and kindly loaned me for examination by Dr. C. P. Blachly—measure 1.08x.75, 1.05x.78.

SUBFAMILY **GALLINULINÆ.** GALLINULES.

GENUS **GALLINULA** BRISSON.

B. 560.   R. 579.   C. 684.   G. ——.   U. 219.

67. Gallinula galeata (LICHT.).   Florida Gallinule.   Prof. F. H. Snow writes me, under date of October 20th, 1885, that since the publication of his "Birds of Kansas," in 1875, he has personally obtained in the State two specimens of *Gallinula galeata*. The first was captured by himself, June 14th, 1878, on the Hackberry, in Gove county; the second by a friend in the vicinity of Lawrence. The bird was entered in his catalogue on the authority of Prof. Baird; and at the time of the publication of my catalogue in 1883, they were known to breed both north and south of the State, and therefore safe to enter as a Kansas bird; but my list only embraced the birds that came under my own observation, and that of others as reported to me. From the fact that the birds nest within their geographical range, and its capture so late in June, I now enter it as a rare summer resident. I have found the birds nesting in Wisconsin as early as the middle of May. Nest in rushes and reeds growing in shallow water, or on swampy lands; build on the tops of old broken-down stalks, and the nests are composed of the same material, weeds, and grasses—also the leaves of the cat-tail flag, when growing in the vicinity — a circular structure, and in some cases quite deep and bulky. Eggs, usually eight to ten; 1.73 x 1.24; buff white, thinly spotted and splashed with varying shades of reddish brown; in form oval. One set of thirteen, collected May 25th, 1878, on a bog in Pewaukee lake, Wisconsin, measure as follows: 1.63x1.18; 1.84x1.27; 1.67x1.18; 1.60x1.16; 1.67x1.18; 1.78x1.30; 1.81x1.29; 1.79x1.29; 1.88x 1.27; 1.70x1.16; 1.80x1.30; 1.75x1.18; 1.80x1.28.

Subfamily FULICINÆ. Coots.

Genus FULICA Linnæus.

B. 559. R. 580. C. 686. G. 271. U. 221.

68. Fulica americana Gmel. American Coot. Summer resident; not uncommon; during migration abundant. Arrive the first to middle of April. Begin laying the last of May. Nest in the tall weeds and rushes growing in shallow, muddy places in ponds and sloughs; built on the tops of the broken-down old growth that forms a platform just above the water; quite a deep, hollow nest, composed of short, bitten-off stems of the weeds and rushes. Eggs, usually eight or nine—I have seen eleven in a nest; 1.92x1.32; cream white, in some cases pale olive drab, thickly and evenly specked with dark brown; in form oval.

Order LIMICOLÆ. Shore Birds.

Family PHALAROPODIDÆ. Phalaropes.

Genus PHALAROPUS Brisson.

Subgenus PHALAROPUS.

B. 520. R. 564. C. 603. G. 264. U. 223.

69. Phalaropus lobatus (Linn.). Northern Phalarope. Migratory; rare. Arrive about the middle to last of May.

Subgenus STEGANOPUS Vieillot.

B. 519. R. 565. C. 602. G. 265. U. 224.

70. Phalaropus tricolor (Vieill.). Wilson's Phalarope. Migratory; common. Arrive the last of April to first of May. June 3d, 1885, I saw a small flock at the edge of a marshy pond in Meade county, and I feel quite confident that they occasionally breed in the western part of the State.

Family RECURVIROSTRIDÆ. Avocets and Stilts.

Genus RECURVIROSTRA Linnæus.

B. 517. R. 566. C. 600. G. 266. U. 225.

71. Recurvirostra americana Gm. American Avocet. Summer resident in western Kansas; rare; during migration common throughout the State. Arrive the last of April to first of May. Begin laying early in June. Nest on the ground in the tall grass at or near the edge of shallow ponds of water; made of the old stems of the grass and lined with the finer leaves of the upland prairie grasses. Eggs, three or four; 1.80x1.30; olivaceous drab to buff, rather uniformly spotted or blotched with varying shades of light to dark brown; in form obovate.

Genus HIMANTOPUS Brisson.

B. 518. R. 567. C. 601. G. —. U. 226.

72. Himantopus mexicanus (Müll.). Black-necked Stilt. Mr. W. H. Gibson, formerly of Topeka, now of Las Vegas, New Mexico, (Taxidermist,) informs me that he saw three of the birds about the middle of June, 1881, on low, wet ground near the Arkansas river at Lakin. Without doubt the birds occasionally breed in southwestern Kansas.

Family SCOLOPACIDÆ. Snipes, Sandpipers, etc.

Genus PHILOHELA Gray.

B. 522.. R. 525. C. 605. G. 240. U. 228.

73. Philohela minor (Gmel.). American Woodcock. Occasional summer resident; quite common in migration. Arrive the last of February to middle of March. Begin laying the first of April. Nest on the ground in the timbered lands along the streams and about the ponds, usually under an old log or at the foot of a stump; a loosely constructed nest of old leaves and grasses. Eggs, three or four; 1.60x1.16; grayish to buff-white, irregularly spotted and blotched with various shades of reddish brown and neutral tints; in shape pyriform to oval.

Genus GALLINAGO Leach.

B. 523. R. 526a. C. 608. G. 241. U. 230.

74. Gallinago delicata (Ord.). Wilson's Snipe. Migratory; common. Arrive in March to first of April.

Genus MACRORHAMPHUS Leach.

B. 525. R. 527a. C. 610. G. 242. U. 232.

75. Macrorhamphus scolopaceus (Say). Long-billed Dowitcher. Migratory; common. Arrive in April.

Genus MICROPALAMA Baird.

B. 536. R. 528. C. 611. G. 243. U. 233.

76. Micropalama himantopus (Bonap.). Stilt Sandpiper. Migratory; rare. Arrive in April.

Genus TRINGA Linnæus.

Subgenus TRINGA.

B. 526. R. 529. C. 626. G. 244. U. 234.

77. Tringa canutus Linn. Knot. Migratory; rare. Two specimens shot in the spring of the year at Neosho Falls, by Col. W. L. Parsons, are the only ones, to my knowledge, captured or seen in the State.

SUBGENUS **ACTODROMAS** KAUP.

B. 531.   R. 534.   C. 616.   G. 245.   U. 239.

78. Tringa maculata VIEILL. Pectoral Sandpiper. Migratory; abundant. Arrive the last of March to middle of April.

B. 533.   R. 536.   C. 617.   G. 246.   U. 240.

79. Tringa fuscicollis VIEILL. White-rumped Sandpiper. Migratory; common. Arrive the last of April to first of May.

B. —.   R. 537.   C. 615.   G. 247.   U. 241.

80. Tringa bairdii (COUES). Baird's Sandpiper. Migratory; quite common. Arrive the last of April to first of May..

B. 532.   R. 538.   C. 614.   G. 248.   U. 242.

81. Tringa minutilla VIEILL. Least Sandpiper. Migratory; abundant. Arrive in April to first of May.

SUBGENUS **PELIDNA** CUVIER.

B. 530.   R. 539*a*.   C. 624.   G. 249.   U. 243*a*.

82. Tringa alpina pacifica (COUES). Red-backed Sandpiper. Migratory; rare. Arrive in April.

GENUS **EREUNETES** ILLIGER. ·

B. 535.   R. 541.   C. 612.   G. 250.   U. 246.

83. Ereunetes pusillus (LINN.). Semipalmated Sandpiper. Migratory; rare. Arrive the last of April to first of May.

GENUS **CALIDRIS** CUVIER.

B. 534.   R. 542.   C. 627.   G. 251.   U. 248.

84. Calidris arenaria (LINN.). Sanderling. Reported by Prof. F. H. Snow in his catalogue of the birds of Kansas: "Migratory; rare. Taken at Lawrence by W. E. Stevens, October 7th, 1874."

GENUS **LIMOSA** BRISSON.

B. 547.   R. 543.   C. 628.   G. 252.   U. 249.

85. Limosa fedoa (LINN.). Marble Godwit. Migratory; not uncommon. Arrive the last of April to first of May.

B. 548.   R. 545.   C. 629.   G. 253.   U. 251.

86. Limosa hæmastica (LINN.). Hudsonian Godwit. Migratory; rare. Arrive the last of April to first of May.

GENUS **TOTANUS** BECHSTEIN.

B. 539.   R. 548.   C. 633.   G. 254.   U. 254.

87. Totanus melanoleucus (GMEL.). Greater Yellow-legs. Migratory; common. Arrive the first of March to first of April.

B. 540.   R. 549.   C. 634.   G. 255.   U. 255.

88. Totanus flavipes (GMEL.). Yellow-legs. Migratory; abundant. Arrive the first of March to first of April.

SUBGENUS **RHYACOPHILUS** KAUP.

B. 541.   R. 550.   C. 637.   G. 256.   U. 256.

89. **Totanus solitarius** (WILS.). Solitary Sandpiper. Migratory; common. Probably breed in the State. Arrive the first of March to first of April.

GENUS **SYMPHEMIA** RAFINESQUE.

B. 537.   R. 552.   C. 632.   G. 257.   U. 258.

90. **Symphemia semipalmata** (GMEL.). Willet. Migratory; rare. Arrive about the first of May. Probably breed in the western part of the State.

GENUS **BARTRAMIA** LESSON.

B. 545.   R. 555.   C. 640.   G. 258.   U. 261.

91. **Bartramia longicauda** (BECHST.). Bartramian Sandpiper. Summer resident; abundant. Arrive the middle of April to first of May. Begin laying early in May. Nest on the prairies in a depression on the ground at the foot of a bunch of grass, and often in open exposed situations; in some cases the bottom of the nest is lined sparingly and loosely with grasses. Eggs, four; 1.75x1.27; grayish white to pale buff, spotted with varying shades of light to dark brown; thickest about large end; in shape pyriform.

GENUS **TRYNGITES** CABANIS.

B. 546.   R. 556.   C. 641.   G. 259.   U. 262.

92. **Tryngites subruficollis** (VIEILL.). Buff-breasted Sandpiper. Migratory; rare. Arrive about the first of May.

GENUS **ACTITIS** ILLIGER.

B. 543.   R. 557.   C. 638.   G. 260.   U. 263.

93. **Actitis macularia** (LINN.). Spotted Sandpiper. Summer resident; rare; in migration common. Arrive the middle of April to first of May. Begin laying the last of May. Nest on the ground, lined sparingly with grasses and leaves; usually on open, dry lands near water, and in a tuft of grass or under a low bush; (I once found a nest under an old drift log). Eggs, four; 1.30x.93; creamy buff to olive drab, spotted and blotched with dark brown and shell markings of lilac; thickest and running somewhat together around large end; in shape pyriform.

GENUS **NUMENIUS** BRISSON.

B. 549.   R. 558.   C. 643.   G. 261.   U. 264.

94. **Numenius longirostris** WILS. Long-billed Curlew. Summer resident; rare; in migration common. Arrive about the first of April. Begin laying early in May. Nest on the ground in a slight depression, sparingly lined with grasses; usually upon the high dry prairies, often quite a distance from water. Eggs, three or four; 2.85x1.85: creamy white to olive drab, spotted and blotched with lilac and varying shades of brown; in form rather oval.

B. 550.   R. 559.   C. 645.   G. 262.   U. 265.

95. **Numenius hudsonicus** LATH. Hudsonian Curlew. Migratory; rare. Arrive the last of April to first of May.

2

B. 551. R. 560. C. 646. G. 263. U. 266.

96. Numenius borealis (Forst.).   Eskimo Curlew.   Migratory; abundant.   Arrive the last of March to middle of April.

Family **CHARADRIIDÆ**. Plovers.

Genus **CHARADRIUS** Linnæus.

Subgenus **SQUATAROLA** Cuvier.

B. 510. R. 513. C. 580. G. 235. U. 270.

97. Charadrius squatarola (Linn.).   Black-bellied Plover.   Migratory; rare. Arrive in April.

Subgenus **CHARADRIUS** Linnæus.

B. 503. R. 515. C. 581. G. 236. U. 272.

98. Charadrius dominicus Müll.   American Golden Plover.   Migratory; abundant.   Arrive about the first of April.

Genus **ÆGIALITIS** Boie.

Subgenus **OXYECHUS** Reichenbach.

B. 504. R. 516. C. 584. G. 237. U. 273.

99. Ægialitis vocifera (Linn.).   Killdeer.   Summer resident; abundant.   Arrive early in the spring; I have often seen the birds in February.   Begin laying the last of April.   Nest on the dry ground, in a small depression, usually beneath a bunch of grass or weeds, in the vicinity of streams and pools of water, lined sparingly with bits of old grass or weeds, chiefly about the edge.   Eggs, usually four; 1.45x1.05; buff to drab white, spotted and blotched with umber and blackish brown, thickest about large end; pyriform in shape, very obtuse at large end and sharply pointed at the other.

Subgenus **ÆGIALITIS** Boie.

B. 507. R. 517. C. 586. G. 238. U. 274.

100. Ægialitis semipalmata Bonap.   Semipalmated Plover.   Migratory; not uncommon.   Dr. Lewis Watson reports seeing a small flock at Ellis, and Mr. V. L. Kellogg and Mr. A. L. Bennett both report seeing several small flocks, and the capture of a pair April 25th, 1885, at Emporia.

Subgenus **PODASOCYS** Coues.

B. 505. R. 523. C. 592. G. 239. U. 281.

101. Ægialitis montana (Towns.).   Mountain Plover.   Summer resident in western to middle Kansas; common.   Arrive about the middle of April.   Begin laying early in May.   Nest in a depression on the ground, lined sparingly with the leaves of grasses.   Eggs, two to four; 1.45x1.10; deep olive to brownish drab, sprinkled with fine dots of blackish brown and neutral tints; pyriform in shape, but not so sharply pointed as other species of this genus.

Order GALLINÆ. Gallinaceous Birds.

Suborder PHASIANI. Pheasants, Grouse, Partridges, Quail, etc.

Family TETRAONIDÆ. Grouse, Partridges, Quail, etc.

Subfamily PERDICINÆ. Partridges.

Genus COLINUS Lesson.

B. 471. R. 480. C. 571. G. 223. U. 289.

102. **Colinus virginianus** (Linn.). Bob-white. Resident; abundant. Begin laying the last of April. Nest in a depression on the ground, usually in the grass upon the prairies, sometimes in a thicket under a low bush; composed of grasses, and generally partially arched over with entrance on the side. Eggs, fifteen to twenty; 1.10x.97; pure white; in shape pyriform.

Subfamily TETRAONINÆ. Grouse.

Genus BONASA Stephens.

B. 465. R. 473. C. 565. G. 218. U. 300.

103. **Bonasa umbellus** (Linn.). Ruffed Grouse. In the early settlement of the State, a resident in eastern Kansas, erroneously, but generally known as "Partridges" in the Northern States, and as "Pheasants" in the Southern States. (The timber along the streams where protected from fire, and the undergrowth from browsing and tramping of cattle, offers a natural home for the birds, and they should be placed there and protected.) Begin laying the last of April. Nest on the ground in groves and at the edge of timber, a place worked out to fit the body, and rather sparingly and loosely lined with grasses and leaves. Eggs, seven to twelve; 1.55x1.15; cream white, occasionally faintly blotched with light drab or buff; in form rather oval, approaching pyriform.

Genus TYMPANUCHUS Gloger.

B. 464. R. 477. C. 563. G. 219. U. 305.

104. **Tympanuchus americanus** (Reich.). Prairie Hen. Resident; common in middle and eastern Kansas. Begin laying the last of April. Nest on the ground in the thick prairie grass, and at the foot of bushes on the barren ground, a hollow scratched out in the soil and sparingly lined with grasses and a few feathers. Eggs, eight to twelve; 1.68x1.25; tawny brown, sometimes with an olive hue, and occasionally sprinkled with brown; in form rather oval.

B. —. R. 477a. C. 564. G. 220. U. 307.

105. Tympanuchus pallidicinctus (RIDGW.). Lesser Prairie Hen. Resident in southern Kansas; rare. Nesting habits similar to *T. americanus.*

GENUS **PEDIOCÆTES** BAIRD.

B. —. R. —. C. —. G. —. U. 308b.

106. Pediocætes phasianellus campestris RIDGW. Prairie Sharp-tailed Grouse. Resident in middle and western Kansas; becoming rare. Begin laying about the middle of May. Nest on the ground, under a bush or tuft of grass on the prairies — a hole scratched out in the earth to fit the body, lined loosely and sparingly with grasses and leaves. Eggs, eight to fifteen; 1.75x 1.25; pale olive drab to rusty brown, usually unmarked, but occasionally uniformly sprinkled with minute dots of dark brown; in form oval. Entered in first catalogue, No. 221, as *P. phasianellus columbianus* (Ord.).

GENUS **CENTROCERCUS** SWAINSON.

B. 462. R. 479. C. 560. G. 222. U. 309.

107. Centrocercus urophasianus (BONAP.). Sage Grouse. Included as an occasional resident of western Kansas on the authority of Mr. Will. T. Cavanaugh, Assistant Secretary of State, who informs me that while hunting buffalo during 1871, 1872, 1873 and 1874, he occasionally met with and shot the birds in the sage brush near the southwest corner of the State. Begin laying the last of May. Nest on the ground upon the plains under a low sage or greasewood bush in a depression scratched out to fit the body; the outside edge or rim composed of small sticks and grasses, lined inside with feathers plucked from their bodies. Eggs, six to nine. In the many nests that I found in Wyoming Territory, seven was the usual number — in no case more than nine; 2.10x1.50; light-greenish drab to pale-yellowish brown, sprinkled with minute dots of reddish brown — upon some a few blotches of reddish brown; in form elongate oval.

FAMILY **PHASIANIDÆ.** PHEASANTS, ETC.

SUBFAMILY **MELEAGRINÆ.** TURKEYS.

GENUS **MELAGRIS** LINNÆUS.

B. 457. R. 470a. C. 554. G. 217. U. 310.

108. Meleagris gallopavo LINN. Wild Turkey. An abundant resident in the early settlement of the State, but rapidly diminishing, and will soon be exterminated. Begin laying early in April. Nest on the ground in dense thickets, often under an old log or tree-top, in a place scratched out to fit the body, and lined loosely and sparingly with grasses, weeds and leaves. Eggs, ten to fifteen; 2.50x1.85; buff white, specked and spotted with rusty brown; in form somewhat oval, but rather pointed at small end, and obtuse at the other.

ORDER COLUMBÆ. PIGEONS.

FAMILY COLUMBIDÆ. PIGEONS.

GENUS ECTOPISTES SWAINSON.

B. 448. R. 459. C. 543. G. 215. U. 315.

109. **Ectopistes migratorius** (LINN.). Passenger Pigeon. Irregular summer resident; rare; a few to my knowledge breed occasionally in the Neosho Valley. Arrive the middle of March. Begin laying about the middle of April. Nest in trees, and in communities; a slight platform structure of twigs, without any material for lining whatever. Eggs, two; 1.45x1.05; white; in form varying from elliptical to oval.

GENUS ZENAIDURA BONAPARTE.

B. 451. R. 460. C. 544. G. 216. U. 316.

110. **Zenaidura macroura** (LINN.). Mourning Dove. Summer resident; abundant; occasional winter sojourner in southern Kansas. Begin laying the last of April. Nest placed on the forks of low, horizontal branches of trees, on grape-vines, and upon the ground; when built off the ground a loose slight platform, constructed of twigs, a few stems of grass and leaves. Eggs, two; 1.12x.85; white; in form elliptical to oval.

ORDER RAPTORES. BIRDS OF PREY.

SUBORDER SARCORHAMPHI. AMERICAN VULTURES.

FAMILY CATHARTIDÆ. AMERICAN VULTURES.

GENUS CATHARTES ILLIGER.

B. 1. R. 454. C. 557. G. 213. U. 325.

111. **Cathartes aura** (LINN.). Turkey Vulture. Summer resident; abundant; occasionally seen in winter. Begin laying the last of April. Nest on rocky ledges, and in hollow trees and stumps. Eggs, two, laid on the bare rocks or debris; no lining; 2.70x1.90; grayish white, variously and unevenly blotched and splashed with light to dark reddish brown and purplish drab; in form rounded oval.

GENUS CATHARISTA VIEILLOT.

B. 3. R. 455. C. 538. G. 214. U. 326.

112. **Catharista atrata** (BARTR.). Black Vulture. Summer resident; rare. Dr. George Lisle, of Chetopa, (a close observer,) writes me in the spring of 1883

that the birds were quite common, and breeding there fifteen or twenty years ago, but now quite scarce; that he saw three of the birds in the fall of 1882 at a "slaughter pen" with Turkey Buzzards; that in 1858 he found a nest with two eggs in an old hollow broken stump. And Dr. Lewis Watson reports the capture of one at Ellis, March 27th, 1885. Nest on the ground, and in old hollow logs and crevices of rocks. Begin laying about the middle of April. Eggs, two, laid on the bare ground and rotten wood, no material of any kind used for lining; 3.00x2.00; dull yellowish to bluish white, spotted and blotched irregularly, in some cases sparingly, on others thickly with umber to dark reddish brown; in form rounded oval.

SUBORDER FALCONES. VULTURES, FALCONS, HAWKS, BUZZARDS, EAGLES, KITES, HARRIERS, ETC.

FAMILY **FALCONIDÆ.** VULTURES, FALCONS, HAWKS, EAGLES, ETC.

SUBFAMILY **ACCIPITRINÆ.** KITES, BUZZARDS, HAWKS, GOSHAWKS, EAGLES, ETC.

GENUS **ELANOIDES** VIEILLOT.

B. 34. R. 426. C. 493. G. 197. U. 327.

113. **Elanoides forficatus** (LINN.). Swallow-tailed Kite. Irregular summer resident in eastern Kansas; some seasons common, others rare. Arrive the first of May. Begin laying the last of May. Nest in the small branches near the tops of tall trees, composed of sticks loosely interwoven, and lined sparingly with the soft, ribbon-like strippings from the inner bark of decaying or dead cottonwood trees. Eggs, four to six; 1.87x1.50; cream white, irregularly spotted and blotched with dark reddish brown, running often largely together towards small end; in form rather oval.

GENUS **ICTINIA** VIEILLOT.

B. 36. R. 428. C. 491. G. 198. U. 329.

114. **Ictinia mississippiensis** (WILS.). Mississippi Kite. Summer resident; rare. Arrive the first of May. Begin laying the last of May. Nest in the forks of trees along the streams, often in deserted crows' nests, fitted up with a few extra sticks and green twigs in leaf, for lining. Eggs, two or three; measurement of one egg collected at Neosho Falls, 1.70x1.35; pure white; in form roundish.

GENUS **CIRCUS** LACEPEDE.

B. 38. R. 430. C. 489. G. 199. U. 331.

115. **Circus hudsonius** (LINN.). Marsh Hawk. Resident; abundant. Begin laying about the first of May. Nest placed on the ground, in the grass, sometimes under low bushes, and usually on the bottom prairie lands; a slight structure, made usually of grasses, sometimes with a foundation of sticks and

weeds. Eggs, four to six; 1.86x1.42; bluish white, generally unspotted, but occasionally with faint to distinct spots and blotches of purplish brown; in form broadly oval.

Genus ACCIPITER Brisson.

Subgenus ACCIPITER.

B. 17. R. 432. C. 494. G. 201. U. 332.

116. Accipiter velox (Wils.). Sharp-shinned Hawk. Winter sojourner; rare. In migration, common.

B. 15, 16. R. 431. C. 495. G. 200. U. 333.

117. Accipiter cooperi (Bonap.). Cooper's Hawk. Resident; common in summer. Begin laying early in May. Nest in the forks of medium-sized trees, from twenty-five to fifty feet from the ground, made of sticks and twigs, and lined sparingly with grass and leaves. Eggs usually four; 1.94x1.54; pale bluish white; occasionally eggs will show faint blotches of lilac to yellowish brown, especially about the large end; in form rounded oval.

Subgenus ASTUR Lacepede.

B. 14. R. 433. C. 496. G. 202. U. 334.

118. Accipiter atricapillus (Wils.). American Goshawk. Winter visitant; rare.

Genus BUTEO Cuvier.

B. 23. R. 436. C. 516. G. 203. U. 337.

119. Buteo borealis (Gmel.). Red-tailed Hawk. Resident; common. Begin laying the last of February. Nest in the forks of the branches of the tallest trees on the timbered bottom lands; a bulky structure, made of sticks and lined sparingly with grass, leaves, and a few feathers. Eggs, three or four; 2.30x1.84; bluish white, thinly and irregularly spotted and blotched with various shades of light to dark brown; in form elliptical to oval.

B. —. R. 436a. C. 519. G. —. U. 337a.

120. Buteo borealis kriderii Hoopes. Krider's Hawk. I killed, October 12th, 1883, a female, near Wallace, and think I saw during the day another, but the birds at a distance so closely resemble the light phase of *Archibuteo ferrugineus* that I was not positive. They are birds of the plains, found from Texas to Minnesota.

B. 20, 24. R. 436b. C. 517. G. 204. U. 337b.

121. Buteo borealis calurus (Cass.). Western Red-tail. Not an uncommon winter sojourner; leave in March.

B. 22. R. 438. C. 515. G. 205. U. 338.

122. Buteo harlani (Aud.). Harlan's Hawk. Winter visitant; rare.

B. 25. R. 439. C. 520. G. 206. U. 339.

123. Buteo lineatus (Gmel.). Red-shouldered Hawk. Resident; common in eastern Kansas. Begin laying early in March. Nest in the forks of branches of medium-sized trees, twenty to fifty feet from the ground, composed of sticks and twigs, and sparingly lined with soft strippings of bark, leaves, and a few feathers. Eggs, three or four; 2.20x1.70; bluish white, irregularly spotted and blotched with varying shades of light to dark-reddish brown; varying in form from subspherical to elliptical.

B. 18, 19, 21, 28.　R. 442.　C. 523.　G. 207.　U. 342.

124. Buteo swainsoni Bonap.　Swainson's Hawk.　Resident; rare in eastern, common in middle and western Kansas.　Begins laying about the middle of May.　Nests vary in height from the shrubby bushes of the plains to the high trees in the timber — a bulky nest constructed of sticks and twigs, and scantily lined with a few weeds or grasses.　Eggs, three to five; 2.25x1.75; dull bluish white; vary greatly in markings, some thinly and rather evenly specked and spotted, others with irregular blotches and splashes of faint to dark-reddish brown, and a few stains of purple; in form rounded oval.

B. 27.　R. 443.　C. 524.　G. 208.　U. 343.

125. Buteo latissimus (Wils.).　Broad-winged Hawk.　A rare bird; probably breeds in eastern Kansas.

Genus **ARCHIBUTEO** Brehm.

B. 30, 31.　R. 447.　C. 525.　G. 209.　U. 347a.

126. Archibuteo lagopus sancti-johannis (Gmel.).　American Rough-legged Hawk.　Winter sojourner; common.　Leave in March.

B. 32.　R. 448.　C. 526.　G. 210.　U. 348.

127. Archibuteo ferrugineus (Licht.).　Ferruginous Rough-leg.　Resident; rare in middle, common in western Kansas.　Begin laying the last of April; nest placed on trees and rocky ledges; a large, bulky structure composed of sticks and twigs, and lined sparingly with weeds or grasses.　Eggs, three or four; 2.50x1.95; buffy white, more or less spotted and blotched with varying shades of light to dark brown; in form rounded oval.

Genus **AQUILA** Brisson.

B. 39.　R. 449.　C. 532.　G. 211.　U. 349.

128. Aquila chrysaetos (Linn.).　Golden Eagle.　Resident; rare.　I am informed by Dr. G. K. Rumsey that a pair nested for several years in the southeastern part of Comanche county, on a high gypsum ledge, and as proof that he was not mistaken, says that the legs of a young bird captured were feathered to the toes.　The late rapid settlement of the county has put a stop to their breeding there.　Begin laying about the middle of March.　Nest usually on the shelves of high, steep rocky cliffs.　May 5th, 1884, I found at Julian, California, a nest placed in and near the top of a tall pine tree — a huge platform structure composed of sticks and twigs, and lined sparingly with grass, moss, and a few feathers.　Eggs, two or three; 2.90x2.25; white, occasionally unmarked, but usually spotted and blotched with various shades of drab to reddish brown, and a few faint purple shell markings; in form broadly subspherical.

Genus **HALIÆETUS** Savigny.

B. 41, 43.　R. 451.　C. 534.　G. 212.　U. 352.

129. Haliæetus leucocephalus (Linn.).　Bald Eagle.　Resident; rare; not uncommon in winter.　The birds in first plumage are dark brown; lighter in color the second year, and generally erroneously known as "Black" and "Gray" Eagles; third year in perfect plumage; that is, with head and tail white.　Only two species in the United States; distinguished in any plumage by the legs.　Tarsi or shank, naked in this species, feathered to the toes in *Aquila* *chrysaëtos.* Begin laying about the middle of March. Nest usually on high trees along the banks of streams, but occasionally, where trees are not convenient, build on high, rocky cliffs, a huge platform structure made of large sticks and lined with twigs, grasses, and a few feathers. Eggs, two to four; average measurements as given by others, about 2.90x2.50; but two that I collected at Neah Bay, Washington Territory, only measure 2.50x1.95, 2.60x2.00; dull white, unmarked; in form rounded oval.

Subfamily FALCONINÆ. Falcons.

Genus FALCO Linnæus.

Subgenus HIEROFALCO Cuvier.

B. 12.   R. 412a.   C. 500.   G. 190.   U. 354.

130. Falco rusticolus Linn. Gray Gyrfalcon. Accidental winter visitant; captured near Manhattan December 1st, 1880, by A. L. Runyan, and reported to me by Dr. C. P. Blachly, who has the bird (a fine specimen) in his collection.

B. 10.   R. 413.   C. 502.   G. 191.   U. 355.

131. Falco mexicanus Schleg. Prairie Falcon. Resident; rare. Nest usually on the side of steep, rocky cliffs, made rudely of sticks and lined with grasses. Eggs are said to be two to four. In Capt. B. F. Goss's collection are two eggs taken April 28th, 1880, at Marysville, Mo., from a tree thirty-five feet from the ground; notes fail to show whether the nest was in the forks of branches or in a hole of the tree, but doubtless in the latter, as the habits of the birds are similar to *Falco peregrinus anatum ;* dimensions of the eggs, 2.05x1.70, 2.12x1.65; grayish white, spotted and blotched with various shades of reddish brown running together so as to obscure the ground color of one of the eggs, and partially of the other; in form rounded oval.

Subgenus RHYNCHODON Nitzsch..

B. 5, 6.   R. 414.   C. 503.   G. 192.   U. 356.

132. Falco peregrinus anatum (Bonap.). Duck Hawk. Resident; not uncommon. Begin laying early in March. Nest in natural cavities in trees and on the sides of rocky cliffs, without lining. Eggs, three or four; 2.25x1.70; grayish ochre, spotted and blotched with reddish and dark chocolate brown, running somewhat together, thickest about large end; in form subspherical to rounded oval.

Subgenus ÆSALON Kaup.

B. 7.   R. 417.   C. 505.   G. 193.   U. 357.

133. Falco columbarius Linn. Pigeon Hawk. Migratory; rare.

B. —.   R. 418.   C. 507.   G. 194.   U. 358.

134. Falco richardsonii Ridgw. Richardson's Merlin. A rare visitant in eastern, not uncommon in western Kansas.

Subgenus TINNUNCULUS Vieillot.

B. 13.   R. 420, 420a.   C. 508, 509.   G. 195.   U. 360.

135. Falco sparverius Linn. American Sparrow Hawk. Resident; abundant. Begin laying the first of April. Nest in large woodpecker holes and natural cavities in the limbs of trees, but little or no material used for lining. Eggs, four to six; 1.33x1.12; buffy white, specked, spotted and blotched with light and dark brown, the markings vary greatly in size and number, often confluent and so thick around large end as to obscure the ground color; in form rounded oval.

SUBFAMILY **PANDIONINÆ.** OSPREYS.

GENUS **PANDION** SAVIGNY.

B. 44.　R. 425.　C. 530.　G. 196.　U. 364.

136. **Pandion haliaetus carolinensis** (GMEL.). American Osprey. Summer resident; rare. Arrive the first of April. Begin laying the last of April. Nest in the tops of trees along the banks and old channels of the rivers; are huge structures, made of large sticks interwoven with corn-stalks and weeds, and lined with grasses. Eggs, two to four; 2.50x1.75; buff white, spotted and blotched with umber and reddish brown running together, thickest at large end; also a few markings of lilac; in form elliptical.

SUBORDER **STRIGES.** OWLS.

FAMILY **STRIGIDÆ.** BARN OWLS.

GENUS **STRIX** LINNÆUS.

B. 47.　R. 394.　C. 461.　G. 181.　U. 365.

137. **Strix pratincola** BONAP. American Barn Owl. Resident; quite common in southern Kansas. Begin laying the last of April. Nest in cavities of trees, burrows in the sides of banks, crevices in rocks, and nooks of buildings. Eggs, four to six; laid upon the debris, pellets of hair, and other remains of the birds' food; 1.70x1.25; cream white; occasionally an egg will show markings of pale drab; in form elliptical.

FAMILY **BUBONIDÆ.** HORNED OWLS, ETC.

GENUS **ASIO** BRISSON.

B. 51.　R. 395.　C. 472.　G. 182.　U. 366.

138. **Asio wilsonianus** (LESS.). American Long-eared Owl. Resident; quite common. Begin laying early in April. Nest in trees and bushes; a coarse, bulky structure, made of sticks, and sparingly lined with grasses, or strips of bark and feathers; often in remodeled hawks' and crows' nests. Eggs, four or five; 1.60x1.30; white; in form subspherical.

B. 52.　R. 396.　C. 473.　G. 183.　U. 367.

139. **Asio accipitrinus** (PALL.). Short-eared Owl. Resident; common. Begin laying the last of April. Nest on the ground, upon the open prairies, generally in the tall grass on the bottom lands, and often beneath a low bush, loosely and slovenly constructed of grass, with occasionally a few leaves and feathers on the bottom. Eggs, four or five; 1.55x1.24; white; in form roundish.

Genus SYRNIUM Savigny.

B. 54. R. 397. C. 476. G. 184. U. 368.

140. Syrnium nebulosum (Forst.). Barred Owl. Resident; common. Begin laying early in March. Nest usually in natural cavities of trees, but have been known to breed in old hawks' nests, upon the branches; a few feathers and leaves constitute the lining. Eggs, three or four; 2.05x1.65; white; in form subspherical.

Genus NYCTALA Brehm.

B. 56, 57. R. 401. C. 483. G. 185. U. 372.

141. Nyctala acadica (Gmel.). Saw-whet Owl. Winter sojourner; rare.

Genus MEGASCOPS Kaup.

B. 49. R. 402. C. 465. G. 186. U. 373.

142. Megascops asio (Linn.). Screech Owl. Resident; abundant. Begin laying early in March. Nest in holes of trees, and occasionally nooks of buildings, lined sparingly with grasses, leaves and feathers. Eggs, four to six; 1.40x1.24; pure white; in form subspherical.

Genus BUBO Cuvier.

B. 48. R. 405. C. 462. G. 187. U. 375.

143. Bubo virginianus (Gmel.). Great Horned Owl. Resident; common. Begin laying the last of February. Nest in natural cavities of trees, deserted nests of hawks, and on the plains or treeless portions of the State in fissures of rocks, scantily lined with leaves and grasses. Eggs, three or four; 2.25x1.90; white; in form subspherical.

B. —. R. 405*a*. C. 463. G. —. U. 375*a*.

144. Bubo virginianus subarcticus (Hoy). Western Horned Owl. October 29th, 1885, I shot a male, in the timber skirting the south fork of Beaver creek, in Rawlins county.

Genus NYCTEA Stephens.

B. 61. R. 406. C. 479. G. 188. U. 376.

145. Nyctea nyctea (Linn.). Snowy Owl. Winter visitant; not uncommon.

Genus SPEOTYTO Gloger.

B. 58, 59. R. 408. C. 487. G. 189. U. 378.

146. Speotyto cunicularia hypogæa (Bonap.). Burrowing Owl. Resident; abundant in middle and western Kansas. Begin laying about the middle of April. Nest in holes in the ground, usually in deserted prairie-dog holes; the end of the burrow is enlarged and lined with grasses, bits of manure, or most any loose soft material at hand. Eggs, four to seven; 1.22x1.04; pure white; in form subspherical.

ORDER PSITTACI. PARROTS, MACAWS, PAROQUETS, ETC.

FAMILY **PSITTACIDÆ.**

GENUS **CONURUS** KUHL.

B. 63. R. 392. C. 460. G. 180. U. 382.

147. Conurus carolinensis (LINN.). Carolina Paroquet. Formerly a common resident in eastern and southern Kansas; but as the settlements increased along the streams, rapidly diminished, and I think have not been met with in the State for several years. In the spring of 1858 a small flock reared their young in a large hollow limb of a giant sycamore tree, on the banks of the Neosho river, near Neosho Falls. I have never been able to procure their eggs; are said to be two or three; greenish white.

ORDER COCCYGES. CUCKOOS, ETC.

SUBORDER **CUCULI.** CUCKOOS, ETC.

FAMILY **CUCULIDÆ.** CUCKOOS, ANIS, ETC.

SUBFAMILY **COCCYGINÆ.** AMERICAN CUCKOOS.

GENUS **GEOCOCCYX** WAGLER.

B. 68. R. 385. C. 427. G. —. U. 385.

148. Geococcyx californianus (LESS.). Road-runner. An occasional visitant in western Kansas. Mr. Charles Dyer, Division Superintendent of the Atchison, Topeka & Santa Fé Railroad at Las Vegas, New Mexico, writes me that in September, 1884, he saw two of the birds near the railroad, and about fifteen miles east of the west line of the State, and that he has seen them quite often in Colorado, near the State line. The birds are known to breed as far east as Las Animas, and I feel confident that they occasionally breed in the south-western corner of the State, a natural habitat for the birds; but unsettled and little known, especially as to its bird-life.

GENUS **COCCYZUS** VIEILLOT.

B. 69. R. 387. C. 429. G. 178. U. 387.

149. Coccyzus americanus (LINN.). Yellow-billed Cuckoo. Summer resident; common. Arrive the first of May. Begin laying the last of May. Nest a loose, light, flat structure, made of sticks and weeds, with at times a little grass for lining, placed in bushes, grape-vines, and on the lower branches of trees, from five to fifteen feet from the ground. Eggs, three to five; 1.25x.90; light bluish green; in form elliptical.

B. 70. R. 388. C. 428. G. 179. U. 388.

150. **Coccyzus erythrophthalmus** (Wils.). Black-billed Cuckoo. A rare summer resident; and not common in migration. Arrive the first of May. A nest was found near Paola in 1863 by Capt. B. F. Goss, and at Manhattan in 1883 by Prof. D. E. Lantz. Nesting habits and eggs (which are a little smaller) similar to *C. americanus*.

Suborder ALCYONES. Kingfishers.

Family ALCEDINIDÆ. Kingfishers.

Genus CERYLE Boie.

Subgenus STREPTOCERYLE Bonaparte.

B. 117. R. 382. C. 423. G. 177. U. 390.

151. **Ceryle alcyon** (Linn.). Belted Kingfisher. Summer resident; common; occasionally lingers into, and I think through, the mild winters. Begin laying about the last of April. Nest at the end of burrows which the birds tunnel horizontally into the sides and near the tops of perpendicular or steep banks of streams, and occasionally into the sides of gravel banks, some distance from the water; are usually about two feet in depth, but have been known to extend over fifteen feet; in fact, not stopping work until a place is reached where they can safely rear their young without fear from falling earth or pebbles. At the end it is scooped out oven-shape, for the nest, which is sometimes sparingly lined with grasses and feathers. Eggs, five or six; 1.32x1.05; pure white; in form oval.

ORDER PICI. WOODPECKERS, WRYNECKS, ETC.

Family PICIDÆ. Woodpeckers.

Genus DRYOBATES Boie.

B. 74. R. 360. C. 438. G. 167. U. 393.

152. **Dryobates villosus** (Linn.). Hairy Woodpecker. Resident; common. Nests excavated in decaying trunks and limbs of trees, or in cavities which it chips into and dresses up to suit. Begin laying the last of April. Eggs, usually four; .96x.73; pure crystal white; in form elliptical.

B. 76. R. 361. C. 440. G. 168. U. 394.

153. **Dryobates pubescens** (Linn.). Downy Woodpecker. Resident; common. Nesting habits similar to *D. villosus*. Begin laying the last of April. Eggs, four or five; .72x.58; pure crystal white; in form elliptical.

GENUS **SPHYRAPICUS** BAIRD.

B. 85.   R. 369.   C. 446.   G. 169.   U. 402.

154. Sphyrapicus varius (LINN.).   Yellow-bellied Sapsucker.   Migratory; rare. Arrive about the middle of April.

B. 86.   R. 369*a*.   C. 447.   G. —.   U. 402*a*.

155. Sphyrapicus varius nuchalis BAIRD.   Red-naped Sapsucker.   Migratory in western Kansas; rare.   I killed a pair out of three young birds found in the willows and cottonwoods thinly skirting the south fork of the Smoky Hill river, at Wallace, October 12th and 14th, 1883.

GENUS **CEOPHLŒUS** CABANIS.

B. 90.   R. 371.   C. 432.   G. 170.   U. 405.

156. Ceophlœus pileatus (LINN.).   Pileated Woodpecker.   Not an uncommon resident along the streams in heavily wooded bottom lands.   Begin laying about the first of April.   Nest in a deep, round hole, chipped out by the bird in a large limb or trunk of a high tree.   Eggs, four or six; average dimensions said to be 1.25x1.00; pure crystal white; in form elliptical.

GENUS **MELANERPES** SWAINSON.

SUBGENUS **MELANERPES.**

B. 94.   R. 375.   C. 453.   G. 172.   U. 406.

157. Melanerpes erythrocephalus (LINN.).   Red-headed Woodpecker.   Summer resident; occasionally linger into the winter.   Begin laying about the middle of May.   Nests in holes which it excavates in trees, telegraph poles, and often, for want of a better place, chips into church steeples and the cornice about the roofs of dwellings.   Eggs, four to six; 1.04x.80; pure translucent white; in form elliptical.

SUBGENUS **ASYNDESMUS** COUES.

B. 96.   R. 376.   C. 456.   G. 158.   U. 408.

158. Melanerpes torquatus (WILS.).   Lewis's Woodpecker.   Reported by Prof. F. H. Snow to the Academy of Science, in the fall of 1878: "Taken at Ellis by Dr. Watson, May 6th, 1878.   One specimen was obtained from a flock of six or eight."

SUBGENUS **CENTURUS** SWAINSON.

B. 91.   R. 372.   C. 450.   G. 171.   U. 409.

159. Melanerpes carolinus (LINN.).   Red-bellied Woodpecker.   Resident; abundant.   Begin laying early in April.   Nesting place excavated in decaying trunks and limbs of trees.   Eggs, four or five; 1.05x.80; pure translucent white; in form elliptical.

GENUS **COLAPTES** SWAINSON.

B. 97.   R. 378.   C. 457.   G. 174.   U. 412.

160. Colaptes auratus (LINN.).   Flicker.   Resident; common.   Begin laying the last of April.   Nests in holes excavated in dead or decaying trunks of trees, and occasionally in church steeples and cornices about buildings.   Eggs, five to seven; 1.03x.84; pure pearly white; in form elliptical.

B. 98a. R. 378a. C. —. G. 175. U. —.

(Colaptes auratus hybridus (Baird) Ridgw. "Hybrid" Flicker. Dropped from the A. O. U. list.)

B. 98. R. 378b. C. 459. G. 176. U. 413.

161. **Colaptes cafer** (Gmel.). Red-shafted Flicker. Resident; rare in eastern Kansas. Begin laying the first of May. Nesting habits and eggs similar to *C. auratus.*

ORDER MACROCHIRES. GOATSUCKERS, SWIFTS, ETC.

Suborder CAPRIMULGI. Goatsuckers, etc.

Family CAPRIMULGIDÆ. Goatsuckers, etc.

Genus ANTROSTOMUS Gould.

B. 112. R. 354. C. 397. G. 163. U. 417.

162. **Antrostomus vociferus** (Wils.). Whip-poor-will. Summer resident; rare; quite common in migration. Arrive the first of May. Begin laying the last of May. Make no nest. Eggs, two, laid in a depression on the ground, among the leaves, in thickets and heavily-wooded lands; are said to average 1.25x.88; one set, all I have to examine, measure only 1.09x.80, 1.12x.78; cream white, irregularly spotted and mottled with lavender and pale brown; in form elliptical.

Genus PHALÆNOPTILUS Ridgway.

B. 113. R. 355. C. 398. G. 164. U. 418.

163. **Phalænoptilus nuttalli** (Aud.). Poor-will. Summer resident; common. Arrive early in May. To be looked for on the high prairies and rocky grounds along the banks of streams. Begin laying the last of May. Eggs, two; 1.05x.80; white, unspotted; laid upon the bare ground, in the thick growths at the edge of the timber; also at the roots of a bunch of bushes or briers, upon the prairies; in form elliptical.

Genus CHORDEILES Swainson.

B. 114. R. 357. C. 399. G. 165. U. 420.

164. **Chordeiles virginianus** (Gmel.). Nighthawk. Summer resident; common in eastern and middle Kansas. Arrive the first to middle of May. Begin laying the last of May. Eggs, two; 1.22x.82; grayish white, thickly mottled all over with varied tints of lilac, purple and yellowish brown; are laid upon the ground on the prairies in bare, open and exposed situations; in form elliptical.

B. 115.   R. 357*a*.   C. 400.   G. 166.   U. 420*a*.

165. Chordeiles virginianus henryi (Cass.). Western Nighthawk. Summer resident in middle and western Kansas; common. Arrive about the middle of May. Begin laying the first of June. Nesting and habits similar to *C. virginianus;* the latter a little smaller and a shade lighter in color.

SUBORDER CYPSELI. SWIFTS.

FAMILY **CYPSELIDÆ.** SWIFTS.

SUBFAMILY **CHÆTURINÆ.** SPINE-TAILED SWIFTS.

GENUS **CHÆTURA** STEPHENS.

B. 109.   R. 351.   C. 405.   G. 162.   U. 423.

166. Chætura pelagica (LINN.). Chimney Swift. Summer resident; abundant in eastern Kansas. Arrive the last of April to first of May. Begin laying the latter part of May. Nests attached to the sides of chimneys and hollow trees, a semi-circular structure made of small sticks of uniform length and size, which are strongly glued together and fastened to the wall with saliva of the birds. Eggs, usually four; .75x.50; pure white, not highly polished; in form long oval.

SUBORDER TROCHILI. HUMMINGBIRDS.

FAMILY **TROCHILIDÆ.** HUMMINGBIRDS.

GENUS **TROCHILUS** LINNÆUS.

SUBGENUS **TROCHILUS.**

B. 101.   R. 335.   C. 409.   G. 161.   U. 428.

167. Trochilus colubris LINN. Ruby-throated Hummingbird. Summer resident; common. Arrive the last of April to first of May. Begin laying the last of May. Nest usually placed on and attached to the top of the body of a small horizontal limb of a tree, six to twelve feet from the ground — a delicate, beautiful nest, composed of a cottony substance, and soft silky fibers from plants, the outside dotted over with lichens. Eggs, two; .48x.33; pure white; in form elliptical.

ORDER PASSERES. PERCHING BIRDS.

SUBORDER CLAMATORES. SONGLESS PERCHING BIRDS.

FAMILY **TYRANNIDÆ.** TYRANT FLYCATCHERS.

GENUS MILVULUS SWAINSON.

B. 123. R. 301. C. 367. G. 149. U. 443.

168. Milvulus forficatus (GMEL.). Scissor-tailed Flycatcher. Summer resident; quite common in southern Kansas. Arrive the first to middle of May. Begin laying the last of May. Nest on the horizontal branches of scrubby trees on and skirting the edge of prairies, six to twelve feet from the ground; a rather flat, loosely constructed nest, composed of sticks, flowering stems of weeds and grasses. Eggs, three to five; .85x.68; white, spotted and blotched with dark red, or reddish brown, and a few purple stains, chiefly about large end; in form rounded oval.

GENUS TYRANNUS CUVIER.

B. 124. R. 304. C. 368. G. 150. U. 444.

169. Tyrannus tyrannus (LINN.). Kingbird. Summer resident; abundant. Arrive the last of April to first of May. Begin laying about the middle of May. Nest in branches of trees, growing in open, exposed situations, six to twelve feet from the ground; a rather bulky, flat structure, composed of stems of weeds and grasses, and lined with hair-like rootlets. I have often found woven in with the same, bits of rags and twine. Eggs, four to six; .90x.68; white, thinly spotted with purple to dark reddish brown; in form oval.

B. 126. R. 306. C. 370. G. 151. U. 447.

170. Tyrannus verticalis SAY. Arkansas Kingbird. Summer resident in middle and western Kansas; common. Arrive about the first of May. Begin laying the last of May. Nesting habits and eggs similar to *T. tyrannus.*

GENUS MYIARCHUS CABANIS.

B. 130. R. 312. C. 373. G. 152. U. 452.

171. Myiarchus crinitus (LINN.). Crested Flycatcher. Summer resident; abundant in eastern Kansas. Arrive the last of April to first of May. Begin laying about the middle of May. Nest in natural cavities of trees, lined with grasses, feathers, hair, and often cast-off skins of snakes. Eggs, four to six; .83x.66; buff white, thickly marked with wavy longitudinal lines, dots and splashes of purple to dark reddish brown; in form oval to elliptical.

3

GENUS **SAYORNIS** BONAPARTE.

B. 135.   R. 315.   C. 379.   G. 153.   U. 456.

172. **Sayornis phœbe** (LATH.). Phœbe. Summer resident; common in eastern Kansas. Arrive in March. Begin laying the last of April. Nest under bridges, overhanging rocks, roots, and suitable places in dwelling and outhouses, composed of layers of mud, moss, grasses, or other miscellaneous material at hand, and warmly lined with fine grasses, rootlets, or hairs. Eggs, four to six; .75x.56; pure white; occasionally sets will be found with dots of reddish brown around large end; in form rounded oval.

B. 136.   R. 316.   C. 377.   G. 154.   U. 457.

173. **Sayornis saya** (BONAP.). Say's Phœbe. Summer resident in western Kansas. Arrive the first of May. A bird of the plains. Begin laying the last of May. Nesting habits and eggs similar to *S. phœbe.*

GENUS **CONTOPUS** CABANIS.

B. 137.   R. 318.   C. 380.   G. 155.   U. 459.

174. **Contopus borealis** (SWAINS.). Olive-sided Flycatcher. Summer resident; rare. Arrive about the middle of May. Begin laying the first of June. Nest in the upper branches of trees, said to be a flat, loose structure, composed of twigs, strips from bark and roots, lined with dry grasses, fragments of moss and lichens. Eggs, three to four; .82x.62; deep cream white, marked around the large end with purple to yellowish and reddish brown; in form oval.

B. 139.   R. 320.   C. 382.   G. 156.   U. 461.

175. **Contopus virens** (LINN.). Wood Pewee. Common summer resident in eastern, rare in western Kansas. Arrive the first of May. Begin laying the last of May. Nest saddled on to the lower limbs of a tree, from eight to twenty-five feet from the ground, composed of fine stemlets, lint-like fibers, rootlets, and bits of cobwebs, the outside coated over with mosses and lichens glued to the material with saliva—a beautiful cup-shaped nest. Eggs, four or five; .73x.52; cream white, spotted and blotched with lilac, purple to dark reddish brown, chiefly at and running together around large end; in form oval.

B. 138.   R. 321.   C. 383.   G. 157.   U. 462.

176. **Contopus richardsonii** (SWAINS.). Western Wood Pewee. Summer resident in western Kansas; rare. Arrive about the middle of May. Begin laying the first of June. Nest usually in the forks or small branches of trees, from eight to thirty feet from the ground, said to be composed chiefly of old dead grasses which are closely woven in and together with fine, linty, thread-like fibers. Eggs, in color, markings and size, similar to *C. virens.*

GENUS **EMPIDONAX** CABANIS.

B. 143.   R. 324.   C. 384.   G. 158.   U. 465.

177. **Empidonax acadicus** (GMEL.). Acadian Flycatcher. Summer resident; not uncommon in eastern Kansas. Arrive the first of May. Begin laying early in June. Nest attached to the forks of branches of trees, six to twelve

feet from the ground, composed chiefly of small, soft, flaxy fibrous strippings from plants; a thin and not very compact structure. Eggs, usually three; .75x.54; cream white, sparingly spotted with reddish brown; in form oval.

B. 140. R. 325a. C. 385. G. 159. U. 466a.

178. Empidonax pusillus traillii (Aud.). Traill's Flycatcher. Migratory; common. Arrive the last of April to first of May. Probably breed in the State.

B. 142. R. 326. C. 387. G. 160. U. 467.

179. Empidonax minimus (Baird). Least Flycatcher. Migratory; not uncommon. Arrive the last of April to first of May.

Suborder OSCINES. Song Birds.

Family ALAUDIDÆ. Larks.

Genus OTOCORIS Bonaparte.

B. 302. R. 300. C. 82. G. 147. U. 474.

180. Otocoris alpestris (Linn.). Horned Lark. Winter sojourner; rare.

B. —. R. —. C. —. G. —. U. 474b.

181. Otocoris alpestris praticola Hensh. Prairie Horned Lark. Resident; common; abundant in eastern Kansas. Begin laying the last of March. Nest in a depression on the ground, under a tuft of grass, made loosely of bits of old grasses, and occasionally lined with hairs. Eggs, four or five; .85x.62; grayish, evenly and thickly spotted with olive brown; in form oval. Entered in first catalogue as *O. alpestris.*

B. —. R. —. C. —. G. —. U. 474c.

182. Otocoris alpestris arenicola Hensh. Desert Horned Lark. Resident; common in middle and western Kansas. Begin laying early in April. Nesting habits and eggs similar to *O. alpestris praticola.* Entered in first catalogue as *O. alpestris leucolæma.*

Family CORVIDÆ. Crows, Jays, Magpies, etc.

Subfamily GARRULINÆ. Magpies and Jays.

Genus PICA Brisson.

B. 432. R. 286. C. 347. G. 145. U. 475.

183. Pica pica hudsonica (Sab.). American Magpie. An occasional fall and winter visitant in western Kansas. Included as a former resident upon the following authority, viz.: Dr. Lewis Watson writes me that Mr. Jeff Jordan reports that while herding cattle in the summer of 1873 or 1874, he found the birds breeding on Brush creek, in Graham county—this was prior to the settlement of the county. From inquiries, I cannot learn that the birds have been seen there since 1875. I called the Doctor's attention to this, and he replies: "I have perfect confidence in Mr. Jordan's statements, and know that he recognizes the birds from his alluding in our late conversation to the two tamed ones that were loose in Ellis last winter." Begin laying early in April. Nest along the streams in low scrubby trees and bushes, from six to fifteen feet from the ground, composed of sticks and twigs, the inside plastered with mud, and lined sparingly with grasses and a few feathers; upon this a rough, dome-like structure of sticks, ingeniously woven, completely covers the nest, leaving a small hole on the side for entrance. Several of the nests that I found in Colorado had two openings, and opposite to each other, doubtless to make room for and protect the long tail of the bird, which must be more or less injured where but one entrance is constructed. Eggs, six to nine; 1.30x.92; light green, thickly specked and spotted with drab to purplish brown; in form broadly oval.

Genus CYANOCITTA Strickland.

B. 434. R. 289. C. 349. G. 146. U. 477.

184. Cyanocitta cristata (Linn.). Blue Jay. Resident; abundant in eastern, common in middle Kansas. Begin laying the last of April. Nest on the branches of trees and bushes in the forests, and in the vicinity of dwellings, six to twenty feet from the ground, composed of sticks and roots strongly interwoven, and lined with rootlets. Eggs, four or five; 1.10x.82; olive, sparingly spotted with drab and olive brown; in form oval.

Subfamily CORVINÆ. Crows.

Genus CORVUS Linnæus.

B. 423, 424. R. 280. C. 338. G. 141. U. 486.

185. Corvus corax sinuatus (Waol.). American Raven. Resident; rare; not uncommon in western Kansas. Begin laying early in April. Nest on the sides of high, precipitous cliffs and in trees — a coarse, bulky structure of sticks, and lined with grasses, hairs, and sometimes bits of wool and moss. Eggs, 1.88x1.30, light greenish-blue, thickly spotted and blotched with purple and blackish brown, in some cases chiefly at large end; in form oval.

B. 425. R. 281. C. 339. G. 142. U. 487.

186. Corvus cryptoleucus (Couch). White-necked Raven. Resident in western Kansas; rare; quite common during the fall and winter. Nesting habits said to be similar to the American Raven.

B. 426. R. 282. C. 340. G. 143. U. 488.

187. Corvus americanus Aud. American Crow. Resident; abundant in eastern but not common in western Kansas. Begin laying the last of March to first of April. Nest in the forks of trees, in groves and on the timbered bottom lands, thirty to seventy-five feet from the ground, composed of sticks, and lined with grasses, hairs, and fibrous strippings from plants and vines. Eggs, four or five; 1.65x1.20; light to dark green, irregularly blotched with purple and dark brown, usually thickest about large end; in form oval.

Genus **CYANOCEPHALUS** Bonaparte.

B. 431.   R. 285.   C. 345.   G. 144.   U. 492.

188. **Cyanocephalus cyanocephalus** (Wied.).   Piñon Jay.   A rare visitant. Three specimens taken October 23d, 1875, near Lawrence, and reported by Prof. F. H. Snow in his catalogue of the birds of Kansas.

Family **ICTERIDÆ.** Blackbirds, Orioles, etc.

Genus **DOLICHONYX** Swainson.

B. 399.   R. 257.   C. 312.   G. 129.   U. 494.

189. **Dolichonyx oryzivorus** (Linn.).   Bobolink.   Summer resident; very rare; during migration quite common.   Arrive the first of May.   Begin laying the last of May; nest in a depression on the ground, in the grass on the low bottom lands, composed of slender, wire-like stems of grasses.   Eggs, four or five; .85x.63; ashy-white, evenly specked with light drab to grayish and reddish brown, and pale surface markings in the shell; in form oval.

Genus **MOLOTHRUS** Swainson.

B. 400.   R. 258.   C. 313.   G. 130.   U. 495.

190. **Molothrus ater** (Bodd.).   Cowbird.   Summer resident; abundant.   Arrive early in March to first of April.   Begin laying about the last of May.   Never build a nest, but drop their eggs into the nests of smaller birds; do not try to take possession by force, but by stealth, during the absence of the owners, and as the birds are polygamous, exhibit no conjugal affections, or love for their offspring.   Average dimensions of their eggs, which vary greatly in size, .85x .65; bluish white, thickly spotted and specked with ashy to reddish brown, and occasionally splashes of purple; in form elliptical.

Genus **XANTHOCEPHALUS** Bonaparte.

B. 404.   R. 260.   C. 319.   G. 131.   U. 497.

191. **Xanthocephalus xanthocephalus** (Bonap.).   Yellow-headed Blackbird. Summer resident; not uncommon; in migration, common.   Arrive the last of April to first of May.   Begin laying the last of May.   Nest in reeds and rushes, composed of flexible leaves of flags and grasses, lined with a finer material of the same, and attached to and woven in and around the standing, growing stalks.   On the first of June I found quite a colony, building in the giant rushes, of the genus Juncus, growing in ponds along Crooked creek, in Meade county; and I have on several occasions found them breeding in small flocks, in different parts of the State.   Eggs, three to six; .95x.74; greenish white, profusely covered with spots and blotches of drab and purplish brown; in form oval.

Genus **AGELAIUS** Vieillot.

B. 401.   R. 261.   C. 316.   G. 132.   U. 498.

192. **Agelaius phœniceus** (Linn.).   Red-winged Blackbird.   Summer resident; abundant.   Arrive in March, a few remaining into and occasionally through

the winter. Begin laying early in May. Nest on low bushes and occasionally in tussocks of grass, on wet marshy grounds, a rather compact basket-like nest, composed of coarse grasses, weeds, and in some cases bits of rushes, fastened to and around the branches upon and against which it rests, and lined with fine grasses. Eggs, four or five; .95x.70; light blue, with thick zigzag markings of light and dark purple and blackish brown around large end, and a few spots of the same colors scattered over the egg; in form oval.

Genus STURNELLA Vieillot.

B. 406. R. 263. C. 320. G. 133. U. 501.

193. Sturnella magna (Linn.). Meadowlark. Resident; abundant in eastern and middle, rare in western Kansas. Begin laying early in May. Nest on the ground in a thick tuft of grass, composed of grasses whic hare often interwoven so as to form a cover overhead. Eggs, four to six; 1.10x.80; white, finely spotted with lilac and reddish brown; in form oval.

B. 407. R. 264. C. 322. G. 134. U. 501b.

194. Sturnella magna neglecta (Aud.). Western Meadowlark. Resident; common in western and middle, rare in eastern Kansas. Begin laying about the middle of May. Nest and eggs similar to the above species.

Genus ICTERUS Brisson.

Subgenus PENDULINUS Vieillot.

B. 414. R. 270. C. 324. G. 135. U. 506.

195. Icterus spurius (Linn.). Orchard Oriole. Summer resident; abundant. Arrive the last of April to first of May. Begin laying the last of May. Nest suspended from twigs, at the end of branches of small trees, along the banks of streams, and in orchards and gardens; a beautiful hemispherical nest, made wholly of a long, slender, wire-like grass, and occasionally bits of a cottony substance, neatly and ingeniously woven together and around the leaf-like twigs that support it. Eggs, four or five; .85x.60; pale, bluish white, thinly marked with specks and zigzag lines of light and bluish brown; thickest about large end; in form oval.

Subgenus YPHANTES Vieillot.

B. 415. R. 271. C. 326. G. 136. U. 507.

196. Icterus galbula (Linn.). Baltimore Oriole. Summer resident; common. Arrive the last of April to first of May. Begin laying the last of May. Nest suspended from the extremities of branches, (the elm appears to be the favorite tree,) fifteen to forty feet from the ground; a compact, strongly-woven, deep, purse-like structure, composed of and attached to the twigs from which it hangs, with flax-like strippings from plants and vines, and lined with hairlike stems of grasses; when in the vicinity of dwellings, twine and thread are used largely in its make-up. Eggs, four or five; .92x.60; pale, bluish white, with a rosy hue when fresh, marked with long, waving lines, and spots of purple and blackish brown, chiefly at large end; in form oblong oval.

B. 416. R. 272. C. 327. G. 137. U. 508.

197. Icterus bullocki (Swains.). Bullock's Oriole. Included on the authority of Prof. F. H. Snow, who enters the same in his catalogue of the birds of Kansas as "summer resident; occurs throughout the State." The birds have not come under my observation in the State.

Genus SCOLECOPHAGUS Swainson.

B. 417.   R. 273.   C. 331.   G. 138.   U. 509.

198. **Scolecophagus carolinus** (Müll.).   Rusty Blackbird.   Winter sojourner; common in eastern Kansas. Leave in March.

B. 418.   R. 274.   C. 331.   G. 139.   U. 510.

199. **Scolecophagus cyanocephalus** (Wagl.).   Brewer's Blackbird.   An occasional summer resident in western Kansas, during migration common; but rare in eastern Kansas. Arrive early in the spring. Begin laying about the 20th of May. The nest is a bulky structure, placed in the forks of trees and bushes, from three to thirty feet from the ground, composed of sticks interlaid with grass, weeds, and tracings of mud, and lined with rootlets and hairs. Eggs, four to six; 1.02x.74; dull greenish white, thickly clouded with specks and blotches of dark reddish brown; in form oval.

Genus QUISCALUS Vieillot.

Subgenus QUISCALUS.

B. —.   R. 278*b*.   C. 336.   G. 140.   U. 511*b*.

200. **Quiscalus quiscula æneus** (Ridgw.).   Bronzed Grackle.   An occasional resident in southern Kansas; abundant in summer. Arrive in March. Begin laying about the middle of April. Nests saddled on to horizontal limbs, or in forks and in excavations of trees, along the streams and in the orchards and shade trees about dwellings; a large and rather compact structure, composed of coarse grasses, weeds, blades of corn, or most any handy material, plastered together with mud and lined with fine grasses, sometimes rootlets and hairs. Eggs, four to six; 1.20x.85; light greenish white, irregularly spotted and marked with zigzag lines of rusty blackish brown, chiefly about large end; in form rounded oval.

Family FRINGILLIDÆ. Finches, Sparrows, etc.

Genus COCCOTHRAUSTES Brisson.

Subgenus HESPERIPHONA Bonaparte.

B. 303.   R. 165.   C. 189.   G. 83.   U. 554.

201. **Coccothraustes vespertina** (Coop.).   Evening Grosbeak.   Migratory; rare.

Genus PINICOLA Vieillot.

B. 304.   R. 166.   C. 190.   G. 84.   U. 515.

202. **Pinicola enucleator** (Linn.).   Pine Grosbeak.   A rare winter visitant.

Genus CARPODACUS Kaup.

B. 305.   R. 168.   C. 194.   G. 85.   U. 517.

203. Carpodacus purpureus (Gmel.).   Purple Finch.   Rare winter sojourner; in migration quite common.   Leave in April.

Genus LOXIA Linnæus.

B. 318.   R. 172.   C. 199.   G. 86.   U. 521.

204. Loxia curvirostra minor (Brehm).   American Crossbill.   Irregular winter visitant; rare.

B. 318a.   R. 172a.   C. 200.   G. ---.   U. 521a.

205. Loxia curvirostra stricklandi Ridgw.   Mexican Crossbill.   On the 13th of November, 1885, Prof. L. L. Dyche, curator of birds and mammals, State University, shot at Lawrence several of the birds out of a small flock, and kindly sent me a pair—*the first capture of the birds in the State.*   On the 21st, Prof. D. E. Lantz reports the killing of three of the birds out of a flock of twelve, at Manhattan; and Mr. V. L. Kellogg a pair out of a flock of twelve, at Emporia, December 23d.

B. 319.   R. 173.   C. 198.   G. 87.   U. 522.

206. Loxia leucoptera Gmel.   White-winged Crossbill.   Irregular winter visitant; rare.

Genus ACANTHIS Bechstein.

B. 320.   R. 179.   C. 207.   G. 88.   U. 528.

207. Acanthis linaria (Linn.).   Redpoll.   Winter visitant; rare.

Genus SPINUS Koch.

B. 313.   R. 181.   C. 213.   G. 89.   U. 529.

208. Spinus tristis (Linn.).   American Goldfinch.   Resident; abundant.   Begin laying late in June.   Nest in the branches of trees and bushes, generally on apple or small elm trees, from six to twelve feet from the ground; constructed of and firmly attached to the limbs upon which it rests with fine hemp-like strippings from plants and bits of cottony substances, and lined with hairs, and now and then a feather.   Eggs, four to six; .65x.50; pale bluish white; when fresh and unblown, with a rosy hue; in form oval.

B. 317.   R. 185.   C. 212.   G. 90.   U. 533.

209. Spinus pinus (Wils.).   Pine Siskin.   Winter sojourner; not uncommon, May 29th, 1883, I shot two birds out of a small flock on the Smoky Hill river, near Wallace.   From their actions I am inclined to think they were breeding there.

Genus PLECTROPHENAX Stejneger.

B. 325.   R. 186.   C. 219.   G. 91.   U. 534.

210. Plectrophenax nivalis (Linn.).   Snowflake.   Winter visitant; rare.

GENUS **CALCARIUS** BECHSTEIN.

B. 326. R. 187. C. 220. G. 92. U. 536.

211. **Calcarius lapponicus** (LINN.). Lapland Longspur. Winter sojourner; abundant. Leave in March.

B. 327. R. 188. C. 221. G. 93. U. 537.

212. **Calcarius pictus** (SWAINS.). Smith's Longspur. Winter sojourner; common in southern Kansas. Leave in March.

B. 328, 329. R. 189. C. 222. G. 94. U. 538.

213. **Calcarius ornatus** (TOWNS.). Chestnut-collared Longspur. Not an uncommon resident in middle and northwestern Kansas; abundant throughout the State in winter. Begin laying early in June. Nest on the ground, on the high open prairie, composed wholly of dry mosses. Eggs, four or five; .72x.56; grayish white, obscurely mottled with pale purple, and this overlaid with spots and splashes of dark reddish brown; in form rather pointed.

GENUS **RHYNCHOPHANES** BAIRD.

B. 330. R. 190. C. 223. G. 95. U. 539.

214. **Rhynchophanes mccownii** (LAWR.). McCown's Longspur. Winter sojourner; quite common in western, rare in eastern Kansas. Leave in March·

GENUS **POOCÆTES** BAIRD.

B. 337. R. 197. C. 232. G. 97. U. 540.

215. **Poocætes gramineus** (GMEL.). Vesper Sparrow. Summer resident; rare; in migration common. Arrive the last of March to first of April. Begin laying early in May. Nest on the ground, made loosely of grasses, and lined with horse hairs. Eggs, four or five; .75x.58; pale greenish white, specked and blotched with various shades of reddish and purple brown; on some the markings are small, chiefly aggregated around the large end; in form oval.

GENUS **AMMODRAMUS** SWAINSON.

SUBGENUS **PASSERCULUS** BONAPARTE.

B. 332. R. 193a. C. 227. G. 96. U. 542a.

216. **Ammodramus sandwichensis savanna** (WILS.). Savanna Sparrow. An occasional winter sojourner in southern Kansas; in migration abundant. Leave in April.

B. 335. R. 193b. C. 229. G. —. U. 542b.

217. **Ammodramus sandwichensis alaudinus** (BONAP.). Western Savanna Sparrow. Migratory. October 14th, 1885, I shot one of the birds, a male, near "Lake Inman," in McPherson county, and saw several others. I am inclined to think they will prove to be quite a common bird in the western part of the State, but they so closely resemble *A. sandwichensis savanna* that they have not been noticed. The birds are, however, considerably smaller, and paler in color. A bleached race of the plains.

SUBGENUS **COTURNICULUS** BONAPARTE.

B. 338.   R. 198.   C. 234.   G. 98.   U. 546.

218. Ammodramus savannarum passerinus (WILS.). Grasshopper Sparrow. Summer resident; abundant. Arrive the last of April to first of May. Begin laying the last of May. Nest on the ground, usually in a tuft of grass, made of old grasses, and sometimes lined with hairs. Eggs, four to six; .75x.60; pure white, thinly spotted with rich reddish brown, thickest about large end; in form rounded oval—almost spherical.

B. 339.   R. 199.   C. 236.   G. 99.   U. 547.

219. Ammodramus henslowii (AUD.). Henslow's Sparrow .Summer resident; rare. Taken in a pasture near Topeka, April 26th, 1872, by Prof. E. A. Popenoe, who has seen the birds since. June 12th I noticed a pair on the high prairies in Woodson county. They nest on the ground, but I have never seen their nests or eggs; are said to differ but little if any from the "Grasshopper Sparrows."

B. 340.   R. 200.   C. 237.   G. 100.   U. 548.

220. Ammodramus leconteii (AUD.). Leconte's Sparrow. Migratory; quite common. Arrive in April.

SUBGENUS **AMMODRAMUS.**

B. ——.   R. 201a.   C. 241.   G. 101.   U. 549a.

221. Ammodramus caudacutus nelsoni ALLEN. Nelson's Sparrow. Summer resident; rare. Arrive the first of May. Nest on the ground. I have never found or seen their nests or eggs.

GENUS **CHONDESTES** SWAINSON.

B. 344.   R. 204.   C. 281.   G. 102.   U. 552.

222. Chondestes grammacus (SAY). Lark Sparrow. Summer resident; abundant. Arrive in April. Begin laying the last of May. Nest usually on the ground; but occasionally in a low tree or bush, composed of branching stems of weeds and grasses, and lined with fine grass, rootlets, and horse-hairs. Eggs, four or five; .80x.65; grayish white, with a few spots and zigzag lines of blackish brown, usually thickest around large end; in form rather rounded.

GENUS **ZONOTRICHIA** SWAINSON.

B. 348.   R. 205.   C. 280.   G. 103.   U. 553.

223. Zonotrichia querula (NUTT.). Harris's Sparrow. Winter sojourner; abundant in southern Kansas. Leave as a whole in March; but a few have been known to occasionally linger until the first of May.

B. 345.   R. 206.   C. 276.   G. 104.   U. 554.

224. Zonotrichia leucophrys (FORST.). White-crowned Sparrow. Migratory; · common. Arrive the last of April to first of May.

B. 346.   R. 207a.   C. 277.   G. —.   U. 555.

225. Zonotrichia intermedia RIDGW. Intermediate Sparrow. Migratory. Quite common in the middle and western part of the State. Arrrive the last of April to first of May.

B. 349. R. 209. C. 275. G. 105. U. 558.

226. Zonotrichia albicollis (GMEL.). White-throated Sparrow. Migratory; common. Arrive the first of April.

GENUS **SPIZELLA** BONAPARTE.

B. 357. R. 210. C. 268. G. 106. U. 559.

227. Spizella monticola (GMEL.). Tree Sparrow. Winter sojourner; abundant. Leave in March.

B. 359. R. 211. C. 269. G. 107. U. 560.

228. Spizella socialis (WILS.). Chipping Sparrow. Summer resident; common. Arrive the last of March to first of April. Begin laying about the middle of May. Prefers to nest about dwellings; always in a low tree or bush, loosely constructed of grass and rootlets, and lined thickly with hairs. Eggs, four or five; .72x.56; bluish green, thinly spotted around large end with purple, light and blackish brown; in form oval.

B. 360. R. 212. C. 272. G. 108. U. 561.

229. Spizella pallida (SWAINS.). Clay-colored Sparrow. Migratory; rare in eastern, common in western Kansas. Arrive the last of April to first of May.

B. 358. R. 214. C. 271. G. 109. U. 563.

230. Spizella pusilla (WILS.). Field Sparrow. Summer resident; quite common in eastern Kansas. Arrive the last of April to first of May. Begin laying about the middle of May. Nest on the ground, also in bushes and low trees; usually upon the ground on uplands, and in trees and bushes on the low bottom lands; loosely constructed of weeds and grasses, and lined with hairs and small threads like stems of plants. Eggs, four or five; .69x.52; grayish white, and as a rule finely and evenly spotted with reddish brown; in form oval.

GENUS **JUNCO** WAGLER.

B. —. R. 216. C. 262. G. 110. U. 566.

231. Junco aikeni RIDGW. White-winged Junco. A single specimen taken at Ellis, by Dr. L. Watson, November 8th, 1875, who now reports that he has seen the birds since on several occasions.

B. 354. R. 217. C. 261. G. 111. U. 567.

232. Junco hyemalis (LINN.). Slate-colored Junco. Winter sojourner; abundant. Leave the first of March.

B. 352. R. 218. C. 263. G. 112. U. 567a.

233. Junco hyemalis oregonus (TOWNS.). Oregon Junco. Winter sojourner; rare in eastern, quite common in middle and western Kansas. Leave the first of March.

GENUS **PEUCÆA** AUDUBON.

B. 371. R. 228. C. 254. G. 113. U. 578.

234. Peucæa cassini (WOODH.). Cassin's Sparrow. Summer resident in middle and western Kansas; not uncommon. Arrive the first to middle of May. Begin laying about the tenth of June. Their favorite resorts and breeding grounds are on the barren plains that are dotted over with low, stunted bushes; are said to nest in bushes and on the ground, and that their eggs are pure white.

GENUS **MELOSPIZA** BAIRD.

B. 363.   R. 231.   C. 244.   G. 114.   U. 581.

235. **Melospiza fasciata** (GMEL.).   Song Sparrow.   Resident in eastern Kansas; rare in summer; common during the winter in the thickets and low sheltered lands.   Begin laying about the first of May.   Nest near the water, usually on the ground, under a tuft of grass, but occasionally in a bush; a compact nest, composed chiefly of grasses, and lined with the slender, hair-like stems.   Eggs, four or five; .78x.59; dull, greenish white, spotted and blotched with reddish brown, and a few purplish stains; the markings are pretty evenly distributed over the entire egg, in some cases sparingly, in others so thick and confluent as to conceal the ground color; in form oval.

B. 368.   R. 234.   C 242.   G. 116.   U. 583.

236. **Melospiza lincolni** (AUD.).   Lincoln's Sparrow.   Migratory; common. Arrive the last of April to first of May.   To be looked for in the timber and brush skirting the streams.   June 16th, 1885, I shot a female in the willows on the bank of the Smoky Hill river, near Wallace.   No signs of the enlargement of the ovaries, and so late in the season, was led to think the bird was breeding there; but the next day, after a faithful search, failed to find either her nest or her mate.

B. 369.   R. 233.   C. 243.   G. 115.   U. 584.

237. **Melospiza georgiana** (LATH.).   Swamp Sparrow.   A rare winter sojourner in eastern Kansas; common during migration.   Leave in April.

GENUS **PASSERELLA** SWAINSON.

B. 374.   R. 235.   C. 282.   G. 117.   U. 585.

238. **Passerella iliaca** (MERB.).   Fox Sparrow.   Winter sojourner; abundant in eastern, rare in western Kansas.   Leave in March to first of April.

B. 376.   R. 235c.   C. 284.   G. 118.   U. 585c.

239. **Passerella iliaca schistacea** (BAIRD).   Slate-colored Sparrow.   Included on the authority of Prof. F. H. Snow, who enters the same in his catalogue of the birds of Kansas, as "migratory; rare."

GENUS **PIPILO** VIEILLOT.

B. 391.   R. 237.   C. 301.   G. 119.   U. 587.

240. **Pipilo erythrophthalmus** (LINN.).   Towhee.   Common resident in eastern Kansas.   Begin laying about the 20th of May.   Nest in thickets and near the edge of timber, usually on the ground, but occasionally on a low tree or bush; a bulky structure composed of leaves, twigs, and strippings from grapevines, and lined with small stems of grasses and rootlets.   Eggs, four or five; .94x.71; grayish white, spotted with reddish brown, thickest and somewhat running together around large end; in form elliptical.

B. 393. R. 238. C. 304. G. 120. U. 588

241. Pipilo maculatus arcticus (Swains.). Arctic Towhee. Winter sojourner; rare in eastern, common in middle and western Kansas. Leave about the first of May.

Genus **CARDINALIS** Bonaparte.

B. 390. R. 242. C. 299. G. 121. U. 593.

242. Cardinalis cardinalis (Linn.). Cardinal. Resident; common in eastern, rare in western Kansas. Begin laying about the middle of May. Nest in low trees, bushes and briers, loosely constructed of leaves, grasses, vine-like stems, and strippings from grape-vines, and lined with finer grasses, which are woven into a rather compact and rounded form. Eggs, three or four; 1.00x.76; grayish white, irregularly spotted with purple, ash and reddish brown, thickest about large end; in form elliptical.

Genus **HABIA** Reichenbach.

B. 390. R. 244. C. 289. G. 122. U. 595.

243. Habia ludoviciana (Linn.). Rose-breasted Grosbeak. Summer resident in eastern Kansas; rare; during migration common. Arrive the first of May. Begin laying the last of May. Nest in small trees in groves and near the edge of timber skirting the streams, six to twelve feet from the ground, generally toward the top and near the center of the tree — a coarse, loosely-constructed nest, made of twigs, stems of weeds, bits of old leaves and rootlets, and lined with a finer material from the same. Eggs, three, occasionally four; .96x.70; greenish white, spotted and blotched with reddish brown; in form oval.

B. 381. R. 245 C. 290. G. 123. U. 596.

244. Habia melanocephala (Swains.). Black-headed Grosbeak. Summer resident in middle and western Kansas; quite common. Arrive first of May. Begin laying the last of May. Nest in low, shrubby trees, on or near the banks of streams, composed of twigs, and stems of weeds or grasses loosely thrown together, and lined with rootlets. Eggs, three or four; 1.00x.68; bluish white, specked and spotted with rusty brown, usually thickest about large end; in form oblong oval.

Genus **GUIRACA** Swainson.

B. 382. R. 246. C. 291. G. 124. U. 597.

245. Guiraca cærulea (Linn.). Blue Grosbeak. Summer resident; quite common in middle and western Kansas. Arrive the first of May. Begin laying the last of May. Nest in bushes and small trees, composed of coarse fibrous strippings, grasses, old leaves, bits of newspapers, and other fragmentary substances, and lined with hairs and rootlets. One taken at Wallace, June 16th, 1885, was built close to the body of a willow tree, on small, twig-like branches, about seven feet from the ground; outside made wholly of narrow strippings of the inner bark of dead cottonwood trees, resting on a foundation of a few old leaves and bits of newspapers, and lined with fine bleached rootlets. Eggs, three or four; .90x.65; bluish white; in form oval.

GENUS **PASSERINA** VIEILLOT.

B. 387.   R. 248.   C. 295.   G. 125.   U. 598.

246. **Passerina cyanea** (LINN.).   Indigo Bunting.   Summer resident; common in eastern Kansas.   Arrive the last of April to first of May.   Begin laying the last of May.   Nest in low bushes, composed of leaves, fibers and grasses, and lined with the finer stems of grasses and horse-hairs.   Eggs, four or five; .75x.58; white, with a faint bluish hue; in form oval.

B. 386.   R. 249.   C. 294.   G. 126.   U. 599.

247. **Passerina amœna** (SAY).   Lazuli Bunting.   Summer resident in western Kansas; rare.   Arrive the first of May.   Begin laying the last of May.   Nest like the "Indigo Bird," in low bushes, and of the same material and make-up. One found May 26th, 1884, in a cañon near San Diego, California, was built near the ends of the branches of a bush, about four feet from the ground, and composed wholly from branching stems of flowering weeds, and lined with the finer stems of the same.   Eggs, usually four; .75x.58; bluish white; in form oval.

B. 384.   R. 251.   C. 292.   G. —.   U. 601.

248. **Passerina ciris** (LINN.).   Painted Bunting.   Summer resident in southwestern Kansas; May 7th to 18th, 1885, I found the birds quite common in the gypsum hills, near the State line.   Arrive the last of April to first of May. Begin laying the last of May.   Nest in the forks of bushes and low trees, composed of grasses, sometimes of leaves at the base, and lined with the finer grasses and hairs.   Eggs, four or five; .70x.53; cream white, thinly specked and spotted with purple and reddish brown, thickest about large end; in form rounded oval.

GENUS **SPIZA** BONAPARTE.

B. 378.   R. 254.   C. 287.   G. 127.   U. 604.

249. **Spiza americana** (GMEL.).   Dickcissel.   Summer resident; abundant in eastern and middle Kansas.   Arrive the first of May.   Begin laying the last of May.   Nest on the ground and in low bushes, usually composed wholly of grasses.   Eggs, four or five; .82x.64; uniform light blue; in form oval.

GENUS **CALAMOSPIZA** BONAPARTE.

B. 377.   R. 256.   C. 286.   G. 128.   U. 605.

250. **Calamospiza melanocorys** STEJN.   Lark Bunting.   Summer resident in middle and western Kansas.   Arrive the last of April to first of May; irregular; some seasons rare, others common.   Begin laying the last of May.   Nest in a depression on the ground, loosely constructed of grasses and stemlets of weeds, and lined with a finer material from the same, and occasionally hairs. Eggs, four or five; .85x.66; light blue; in form rounded oval.

GENUS **PIRANGA** VIEILLOT.

B. 220.   R. 161.   C. 154.   G. 81.   U. 608.

251. **Piranga erythromelas** VIEILL.   Scarlet Tanager.   Summer resident; common in eastern Kansas.   Arrive the last of April.   Begin laying about the

20th of May. Nest on horizontal limbs of trees, (occasionally near houses, but as a rule a retiring bird, and an inhabitant of the forest,) a flat and loosely constructed nest of stems and strips from plants, and lined with fine, hair-like fibers and rootlets. Eggs, four or five; .90x.65; pale greenish blue, minutely spotted with reddish brown, and occasional markings of obscure purple, often aggregated into a wreath around large end; in form oval.

B. 221. R. 164. C. 155. G. 82. U. 610.

252. **Piranga rubra** (LINN.). Summer Tanager. Summer resident; common in eastern Kansas. Arrive the last of April. Begin laying about the 20th of May. Nesting habits similar to those of the Scarlet Tanager, but not so retiring. Eggs, three or four; .90x68; light emerald green, specked and spotted with various shades of purple and dark brown; thickest and running together around large end; in form oval.

FAMILY **HIRUNDINIDÆ.** SWALLOWS.

GENUS **PROGNE** BOIE.

B. 231. R. 152. C. 165. G. 75. U. 611.

253. **Progne subis** (LINN.). Purple Martin. Summer resident; common. First arrivals the last of March. Begin laying about the last of April. Nest in boxes and gourds erected for them; also in deserted woodpecker holes and cavities in trees; composed of various materials loosely thrown together, such as dry grasses, straws, bits of strings, etc., and warmly lined with feathers. Eggs, four to six; .97x.68; cream white; in form oval.

GENUS **PETROCHELIDON** CABANIS.

B. 226. R. 153. C. 162. G. 76. U. 612.

254. **Petrochelidon lunifrons** (SAY). Cliff Swallow. Summer resident; abundant. Arrive the first of May. Begin laying about the middle of May. Nest in communities against the side of vertical rocky cliffs and under the eaves of buildings; composed of mud, and lined with dry grasses, leaves, and feathers; when in exposed positions are in the shape of a retort, with entrance to passage-way from beneath; but under the eaves or in sheltered places are more globular, and without the long necks. Eggs, usually four; .77x.56; white, dotted and blotched with dark-reddish brown; the markings varying greatly in size, number, and distribution; in form rounded oval.

GENUS **CHELIDON** FORSTER.

B. 225. R. 154. C. 159. G. 77. U. 613.

255. **Chelidon erythrogaster** (BODD.). Barn Swallow. Summer resident; common. Arrive in April. Begin laying about the middle of May. Nest attached to outbuildings, the sides of rafters in the barn, and in unsettled portions of the country under overhanging rocks and in cave-like cavities; constructed of layers of mud and grasses, and lined with fine grasses and downy feathers. Eggs, four to six; .78x.55; white, with spots and blotches of dark-reddish brown and purple, chiefly about large end; in form long oval.

Genus **TACHYCINETA** Cabanis.

B. 227.   R. 155.   C. 160.   G. 78.   U. 614.

256. Tachycineta bicolor (Vieill.).   Tree Swallow.   Summer resident; rare; in migration common.   Arrive the last of March.   Begin laying the first of May.   Nest in deserted Woodpecker holes, natural cavities in the trees, and occasionally in boxes — (I once found one in a fence rail;) is loosely constructed of fine dry grasses and leaves, and thickly lined with downy feathers. Eggs, four to six; .78x.52; pure transparent white; in form rather pointed oval.

Genus **CLIVICOLA** Forster.

B. 229.   R. 157.   C. 163.   G. 79.   U. 616.

257. Clivicola riparia (Linn.).   Bank Swallow.   Summer resident; common. Arrive in April.   Begin laying about the middle of May.   Nest in communities, in holes made by themselves on the sides of perpendicular banks, usually near the top, and about three feet in depth, but in gravelly soil have been known to go great distances, or until a place free from stones overhead has been reached, (this is evidently to prevent injury to their eggs, or young, from falling earth or pebbles,) the end worked out oven-shape, and lined with fine grasses and feathers.   Eggs, four to six; .68x.45; pure white; when unblown have a rosy hue; in form oval.

Genus **STELGIDOPTERYX** Baird.

B. 230.   R. 158.   C. 164.   G. 80.   U. 617.

258. Stelgidopteryx serripennis (Aud.).   Rough-winged Swallow.   Summer resident; common.   Arrive in April.   Begin laying about the middle of May. Nest in holes in banks of streams, constructed of the same material as the Barn Swallow, but not so deep or uniform, and often in crevices, old excavations, and openings from various causes.   Eggs, five or six; .70x.52; pure white; in form rather long and pointed.

Family **AMPELIDÆ.** Waxwings, etc.

Subfamily **AMPELINÆ.** Waxwings.

Genus **AMPELIS** Linnæus.

B. 272.   R. 150.   C. 166.   G. 73.   U. 618.

259. Ampelis garrulus Linn.   Bohemian Waxwing.   Winter visitant; very rare.

B. 233.   R. 151.   C. 167.   G. 74.   U. 619.

260. Ampelis cedrorum (Vieill.).   Cedar Waxwing.   Resident; irregular; some years abundant, others rare.   Begin laying about the 20th of June. Nest built in apple and other low trees, sometimes in bushes; a bulky structure, composed of twigs, stems of weeds, grasses, and coarse fibrous strippings from vines and plants, lining the same sparingly with leaves and fine rootlets.   Eggs, four to six; .85x.60; pale clay-white, with an olive hue, thinly spotted with purple, and light to very dark brown; in form oval.

Family **LANIIDÆ.** Shrikes.

Genus **LANIUS** Linnæus.

B. 236. R. 148. C. 186. G. 70. U. 621.

261. Lanius borealis Vieill. Northern Shrike. Winter sojourner; quite common. Leave in March.

B. 237. R. 149. C. 187. G. 71. U. 622.

262. Lanius ludovicianus Linn. Loggerhead Shrike. Prof. F. H. Snow, in his catalogue of the birds of Kansas, says: "Several typical specimens of this southern form have been taken." The birds so far have not come under my observation in the State.

B. 238. R. 149a. C. 188. G. 72. U. 622a.

263. Lanius ludovicianus excubitorides (Swains.). White-rumped Shrike. Summer resident; common; occasionally linger into winter. Arrive early in the spring. Begin laying early in May. Nest in thorn trees, hedges and briers, composed of small sticks and stems, with bits of leaves, wood, feathers, and other soft fragmentary substances, sparingly woven in, lined with fine stems of weeds and grasses, and in some cases hairs. Eggs, four to six; 1.02x.73; dull, yellowish white, spotted and blotched with ash-purple and brown, more or less confluent, thickest around large end; in form oval.

Family **VIREONIDÆ.** Vireos.

Genus **VIREO** Vieillot.

Subgenus **VIREOSYLVA** Bonaparte.

B. 240. R. 135. C. 170. G. 64. ·U. 624.

264. Vireo olivaceus (Linn.). Red-eyed Vireo. Summer resident; abundant. Arrive the last of April to first of May. Begin laying about the 20th of May. Nest pensile, suspended from the forks or twigs of forest trees, made of and fastened to and around the limbs with lint-like fibers, shreds from weeds, vines, bits of old leaves, spider-threads and cocoons, woven in and fastened together with saliva, and lined with hair-like stems and rootlets; to be looked for anywhere from the lowest branches to near the tops of the tallest trees. Eggs, three to five; .80x.56; pure white, thinly and irregularly specked with reddish brown, chiefly at large end; in form oval.

B. 245. R. 139, 139a. C. 174, 175. G. 65. U. 627.

265. Vireo gilvus (Vieill.). Warbling Vireo. Summer resident; common. Arrive the last of April to first of May. Begin laying about the 20th of May. Nest of about the same material and in similar positions to the "Red-eyed," but not so retiring; generally build in the vicinity of dwellings, and in make-up more smooth and compact. Eggs, four or five; .75x.55; crystal white, sparingly spotted at the larger end with light and dark brown; in form oval.

4

SUBGENUS **LANIVIREO** BAIRD.

B. 252.   R. 140.   C. 176.   G. 66.   U. 628.

266. Vireo flavifrons VIEILL.   Yellow-throated Vireo.   Summer resident; quite common.   Arrive the last of April to first of May.   Begin laying about the 20th of May.   Nest a pendent one, like all of the Vireo family, but readily distinguished from the others by being thickly adorned on the outside with lichens, never very high from the ground, and in rather an open and exposed situation.   All that I have found were in the timber and away from settlements, but writers in the Eastern States speak of them as a familiar bird, nesting in orchards and in gardens, which I have no doubt will be the case here wherever the trees and shrubbery around our prairie homes form an inviting haunt. Eggs, four or five; .82x.58; white, with a very few scattering spots, chiefly at large end, of dark rosy brown; in form oval.

B. 250.   R. 141.   C. 177.   G. 67.   U. 629.

267. Vireo solitarius (WILS.).   Blue-headed Vireo.   Migratory; rare.   Arrive first of May.

SUBGENUS **VIREO** VIEILLOT.

B. 247.   R. 142.   C. 185.   G. —.   U. 630.

268. Vireo atricapillus WOODH.   Black-capped Vireo.   Summer resident in the gypsum hills in southwestern Kansas.   The habits of the birds are but little known.   On the 11th of May, 1885, I found the birds building a nest near the head of a deep cañon, suspended from the forks of a small elm tree about five feet from the ground, hemispherical in shape, and composed of broken fragments of bleached leaves, with here and there an occasional spider's cocoon, interwoven with and fastened to the twigs with fibrous strippings, threads from plants, and the webs of spiders, and lined with fine stems from weeds and grasses; above, it was screened from sight by the thick foliage of the trees, but beneath, for quite a distance, there was nothing to hide it from view.   The material of which it was made, however, so closely resembled the gypsum that had crumbled from the rocks above, that the casual observer would have passed it by unnoticed.   I regret that I could not stay for the eggs, but as the birds are quite common in that vicinity (southeastern Comanche county), I trust that before another season passes I shall be able to describe the eggs.

B. 248.   R. 143.   C. 181.   G. 68.   U. 631.

269. Vireo noveboracensis (GMEL.).   White-eyed Vireo.   Summer resident; common.   Arrive the last of April to first of May.   Begin laying about the middle of May.   Nest on low bottom lands at the edge of the timber, in thickets, suspended in a small open space from grape and other wild running vines and briers, two to four feet from the ground; made of hemp-like fibers, bits of old leaves, and mosses from decaying stumps and logs, and lined with fine stemlets of weeds and grasses.   Eggs, four or five; .73x.54; clear white, with a few scattering spots of purple and dark-reddish brown about large end; in form oval.

B. 246.   R. 145.   C. 183.   G. 69.   U. 633.

270. Vireo bellii AUD.   Bell's Vireo.   Summer resident; abundant.   Arrive the last of April to first of May.   Begin laying the last of May.   Nest in hedges, vines and small thickets on or at the edge of the prairies, suspended with and composed of fibrous, lint-like strippings from plants, interwoven with bits of old leaves and other fragmentary substances, and lined with fine, slender stemlets from weeds and grasses; in some cases lined with hairs. Eggs, four; .70x.51; pure white, thinly specked or dotted around large end with darkreddish brown; in form oval.

FAMILY **MNIOTILTIDÆ.** WOOD-WARBLERS.

GENUS **MNIOTILTA** VIEILLOT.

B. 167a. R. 74, 74a. C. 91, 92. G. 32. U. 636.

271. Mniotilta varia (LINN.). Black and White Warbler. Summer resident; quite common in eastern Kansas. Arrive the last of April to first of May. Begin laying about the 20th of May. Nest on the ground, composed of strippings of plants, grasses, moss, leaves, and the inner bark from decaying trees, and lined with fine stems of grass and hairs; occasionally partially roofed over. Eggs, from four to seven; measurement of a set of five: .66x.52, .66x.52, .65x.51, .64x.52, .64x.50; white, speckled with umber and reddish brown, chiefly at large end; in some cases a few purplish spots; in form oval.

GENUS **PROTONOTARIA** BAIRD.

B. 169. R. 75. C. 95. G. 33. U. 637.

272. Protonotaria citrea (BODD.). Prothonotary Warbler. Summer resident; common in eastern Kansas. Arrive about the first of May. Begin laying the last of May. Nest in Woodpecker holes, openings or niches in trees, stumps, and outbuildings, on the banks of streams and ponds, never far from the ground, composed of moss, grasses, dry leaves, lichens and even bits of rotten wood interwoven with fine rootlets, and lined with hair. Eggs, four to seven; .68x.56; cream white, thickly spotted with lilac, purple and dark brown, thickest and often confluent at large end; in form elliptical.

GENUS **HELMITHERUS** RAFINESQUE.

B. 178. R. 77. C. 96. G. 34. U. 639.

273. Helmitherus vermivorus (GMEL.). Worm-eating Warbler. Migratory; rare. Probably breed in the State. Arrive the last of April to first of May.

GENUS **HELMINTHOPHILA** RIDGWAY.

B. 180. R. 79. C. 98. G. 35. U. 641.

274. Helminthophila pinus (LINN.). Blue-winged Warbler. Summer resident; rare; in migration common. Arrive the last of April to first of May. Nest on the ground, generally at the edge of low thickets. I have never found or seen their nest. Audubon says: "It is singularly constructed, and of an elongate, inversely conical form; is attached to several stalks or blades of tall grass by its upper edge. The materials of which it is formed are placed obliquely from its mouth to the bottom. The latter part is composed of dried leaves, and is finished within with fine grass and lichens." Eggs, four or five; .66x.50; white, thinly specked with reddish brown, chiefly at large end; in form oval.

B. 183. R. 85. C. 106. G. 36. U. 645.

275. Helminthophila ruficapilla (WILS.). Nashville Warbler. Migratory; rare. Arrive the last of April to first of May.

B. 184. R. 86. C. 107. G. 37. U. 646.

276. Helminthophila celata (SAY). Orange-crowned Warbler. Migratory; common. Arrive the last of April to first of May.

B. 185. R. 87. C. 109. G. 38. U. 647.

277. Helminthophila peregrina (WILS.). Tennessee Warbler. Migratory; common. Arrive the last of April to first of May.

GENUS **COMPSOTHLYPIS** CABANIS.

B. 168. R. 88. C. 93. G. 39. U. 648.

278. Compsothlypis americana (LINN.). Parula Warbler. Migratory; common. Arrive the last of April to first of May. I noticed a pair in the latter part of July, 1879, feeding young birds in the tree-tops, near Neosho Falls; in flight not strong enough to have come far, and I am inclined to think the birds occasionally breed in the State.

GENUS **DENDROICA** GRAY.

SUBGENUS **DENDROICA** GRAY.

B. 203. R. 93. C. 111. G. 40. U. 652.

279. Dendroica æstiva (GMEL.). Yellow Warbler. Summer resident; abundant. Arrive in April. Begin laying about the middle of May. Nest in small trees and bushes, giving preference to orchards and shrubbery in gardens, a neatly constructed nest of fibrous strippings, and a cotton-like substance from plants, and lined sparingly with fine grasses, hairs, and now and then a feather. Eggs, four or five; .65x.50; bluish white, with specks and blotches of brown, umber and lilac, irregularly scattered over the eggs, thickest around large end; in form oval.

B. 194. R. 95. C. 110. G. 41. U. 655.

280. Dendroica coronata (LINN.). Myrtle Warbler. An occasional winter sojourner; in migration abundant. Leave in April.

B. 195. R. 96. C. 120. G. 42. U. 656.

281. Dendroica auduboni (TOWNS.). Audubon's Warbler. In western Kansas; migratory; not uncommon. Arrive early in the spring.

B. 204. R. 97. C. 125. G. 43. U. 657.

282. Dendroica maculosa (GMEL.). Magnolia Warbler. Migratory; rare. Arrive the first of May.

B. 201. R. 98. C. 118. G. 44. U. 658.

283. Dendroica cærulea (WILS.). Cerulean Warbler. Summer resident; rare; in migration common. Arrive the first of May. Inhabit and nest in the tree tops on the timbered bottom lands. Begin laying the last of May. Nest usually placed on the forks of small branches, twenty to sixty feet from the

ground, composed of fine grasses, moss, and bits of hornets' nests, interwoven
with spider-webs, and twined with soft fine strippings from plants, the outside sparingly dotted with lichens. Eggs, four or five; .60x.47; cream white,
with a few reddish spots, chiefly at large end; in form oval.

B. 200.   R. 99.   C. 124.   G. 45.   U. 659.

284. Dendroica pensylvanica (LINN.).   Chestnut-sided Warbler.   Migratory;
rare.   Taken at Leavenworth in May, 1871, by Prof. J. A. Allen, and near Topeka, May 2d, 1873, by Prof. E. A. Popenoe.   May nest in the State.

B. 202.   R. 101.   C. 122.   G. 46.   U. 661.

285. Dendroica striata (FORST.).   Black-poll Warbler.   Migratory; common.
Arrive the first of May.

B. 196.   R. 102.   C. 121.   G. 47.   U. 662.

286. Dendroica blackburniæ (GMEL.).   Blackburnian Warbler.   Migratory;
very rare.   Specimen shot at Leavenworth, May 4th, 1881, by Prof. J. A.
Allen.

B. —.   R. 103a.   C. 130.   G. 48.   U. 663a.

287. Dendroica dominica albilora BAIRD.   Sycamore Warbler.   Summer resident; rare.   Arrive the last of April to first of May.   Begin laying the last
of May.   Nest in the trees and bushes, composed of mosses and lichens, and
lined with fine, soft, fibrous strippings from plants.   Eggs, four; .70x.52;
white, spotted with purple and brown; aggregating at and often forming a
wreath around large end; in form oval.

B. 189.   R. 107.   C. 112.   G. 49.   U. 667.

288. Dendroica virens (GMEL.).   Black-throated Green Warbler.   Migratory;
rare.   Arrive the first of May.

B. 198.   R. 111.   C. 134.   G. 50.   U. 671.

289. Dendroica vigorsii (AUD.).   Pine Warbler.   Migratory; rare.   Arrive the
first of May.   Probably breed in the State.

B. 208.   R. 113.   C. 132.   G. 51.   U. 672.

290. Dendroica palmarum (GMEL.).   Palm Warbler.   Migratory; rare.   Arrive
the first of May.

B. 210.   R. 114.   C. 127.   G. 52.   U. 673.

291. Dendroica discolor (VIEILL.).   Prairie Warbler.   Summer resident; rare.
Arrive the last of April to first of May.   Begin laying the last of May.   Nest
in bushes and on lower branches of trees in open or thinly-wooded lands, two
to eight feet from the ground, placed in upright forks of the twig-like branches,
made of leaves and strippings from plants, and lined with hair-like rootlets.
Eggs, four or five; .67x.49; white, thinly spotted with lilac, purple and brown;
in form oval.

GENUS **SEIURUS** SWAINSON.

B. 186.   R. 115.   C. 135.   G. 53.   U. 674.

292. Seiurus aurocapillus (LINN.).   Oven-bird.   Summer resident; common.
Arrive about the first of May.   Begin laying about the middle of May.   Nest
on the ground, generally a depression among the leaves, and hidden under a
low bush, log or overhanging roots; when in an open space, roofed over; a

dome-like structure made of leaves and grasses, with entrance on the side. Eggs, four or five; .80x.55; white, marked around large end with dots and blotches of reddish brown and lilac; in form rounded oval.

B. 187. R. 116. C. 136. G. 54. U. 675.

293. **Seiurus noveboracensis** (GMEL.). Water-Thrush. Migratory; rare. Arrive the last of April to first of May. Possibly breed in the northern part of the State.

B. 188. R. 117. C. 138. G. 55. U. 676.

294. **Seiurus motacilla** (VIEILL.). Louisiana Water-Thrush. Summer resident; common. Arrive the last of April. Begin laying about the 8th of May. Nest on the ground, under projecting roots, old logs, and fissures in rocks, on the banks of streams and ponds, and near the water's edge, composed of leaves and mosses, and lined with fine grasses, fibers and hairs. Eggs, four or five; .78x.59; white, specked with reddish brown, thickest around large end; in form oval.

GENUS **GEOTHLYPIS** CABANIS.

SUBGENUS **OPORORNIS** BAIRD.

B. 175. R. 119. C. 140. G. 56. U. 677.

295. **Geothlypis formosa** (WILS.). Kentucky Warbler. Summer resident; common. Arrive the last of April. Begin laying about the 20th of May. Nest on the ground, usually on the banks of streams, in thick growths of small trees; outside or base a loose structure of leaves, stems, and wide blades of grass, upon which a more compact inner nest is built of the finer grasses, stems and rootlets, and lined with horse hair. Eggs, four or five; .72x.53; white, finely dotted with reddish brown, chiefly around large end; in form oval.

SUBGENUS **GEOTHLYPIS** CABANIS.

B. 172. R. 120. C. 142. G. 57. U. 679.

296. **Geothlypis philadelphia** (WILS.). Mourning Warbler. Migratory; rare. Arrive the last of April.

B. —. R. —. C. —. G. —. U. 681a.

297. **Geothlypis trichas occidentalis** BREWST. Western Yellow-throat. Summer resident; abundant. Arrive in April. Begin laying about the 20th of May. Nest usually on the ground, but I have found them in bushes, two or three feet from the ground, composed outside loosely of leaves and grasses, inside of wire-like stems from plants and rootlets interwoven together. Eggs, four to six; .68x.50; clear white, spotted and blotched with reddish brown and purple, thickest around large end; in form oval. Entered in first catalogue (No. 58) as *G. trichas*.

GENUS **ICTERIA** VIEILLOT.

B. 176. R. 123. C. 144. G. 59. U. 683.

298. **Icteria virens** (LINN.). Yellow-breasted Chat. Summer resident; common. Arrive the first of May. Begin laying about the middle of May. Nest generally in thickets, on low bushes; outside composed of leaves, within with layers of strippings from the bark of grape-vines and weeds, lined with fine grasses and fibrous roots. Eggs, four or five; .85x.66; glossy white; finely spotted with a rich reddish brown, thickest about large end; in form elliptical.

B. 177. R. 123*a*. C. 145. G. ——. U. 683*a*.

299. Icteria virens longicauda (Lawu.). Long-tailed Chat. A summer resident in the western part of the State; not uncommon. Nesting habits, eggs and actions are similar to those of the Yellow-breasted; but note and song slightly different. The birds were reported by Prof. F. H. Snow, in vol. 6, page 38, Transactions of the Kansas Academy of Science, as "Taken along the Smoky Hill river in western Kansas by S. W. Williston, in May, 1877;" but by oversight omitted from first catalogue. Attention was immediately called to the same. (See Bulletin of the Nuttall Ornithological Club, vol. 8, page 227.) June 2, 1885. I shot two of the birds on Crooked creek, in Meade county, and saw several others.

GENUS **SYLVANIA** NUTTALL.

B. 211. R. 124. C. 146. G. 60. U. 684.

300. Sylvania mitrata (GMEL.). Hooded Warbler. A summer resident in eastern Kansas; rare. Arrive about the first of May. Begin laying the last of May. Nest in the forks of low bushes on bottom and marshy lands, composed of leaves, strippings from plants, grasses, and a cotton-like substance, and lined with fine, hair-like stems. Eggs, four or five; .68x.50; white, spotted around large end with reddish brown and a few purplish stains; in form oval.

B. 213. R. 125. C. 147 G. 61. U. 685.

301. Sylvania pusilla (WILS.). Wilson's Warbler. Migratory; quite common. Arrive the last of April to first of May.

B. 214, 215. R. 127. C. 149. G. 62. U. 686.

302. Sylvania canadensis (LINN.). Canadian Warbler. Migratory; rare. Arrive the last of April to first of May.

GENUS **SETOPHAGA** SWAINSON.

B. 217. R. 128. C. 152. G. 63. U. 687.

303. Setophaga ruticilla (LINN.). American Redstart. Summer resident; common; in migration abundant. Arrive the last of April to first of May. Begin laying the last of May. Nest in small trees, usually six to ten feet from the ground, (but I have found their nests all the way from three to thirty feet from the ground,) usually placed within and woven around three or more small upright branches, composed of stems, rootlets, strippings from plants, and a soft, fibrous, cottony substance, which is worked in and covers the outside; the inside is lined with fine stems, hairs, and occasionally a few feathers; a neat, compact structure. Eggs, four; .67x.50; cream white, dotted with fine specks of reddish brown and lilac, thickest and running together around large end; in form oval.

FAMILY **MOTACILLIDÆ.** WAGTAILS.

GENUS **ANTHUS** BECHSTEIN.

SUBGENUS **ANTHUS.**

B. 165. R. 71. C. 89. G. 30. U. 697.

304. Anthus pensilvanicus (LATH.). American Pipit. Migratory; quite common. Arrive about the first of April.

SUBGENUS **NEOCORYS** SCLATER.

B. 166.   R. 73.   C. 90.   G. 31.   U. 700.

305. Anthus spragueii (AUD.) Sprague's Pipit. Migratory; rare. Arrive about the first of April.

FAMILY **TROGLODYTIDÆ**. WRENS, THRASHERS, ETC.

SUBFAMILY **MIMINÆ**. THRASHERS.

GENUS **MIMUS** BOIE.

B. 253, 253a.   R. 11.   C. 15.   G. 7.   U. 703.

306. Mimus polyglottos (LINN.). Mocking Bird. Summer resident; becoming quite common. Arrive the middle of April to first of May. Begin laying about the 20th of May. Nest in small trees, thickets, hedges, and in various locations, often near houses; but rarely over ten feet from the ground. Outside loosely constructed of sticks and weeds, and lined with fine rootlets. Eggs, four or five; .99x.75; light-greenish blue, spotted and blotched with yellowish to very dark brown and purple; thickest about large end; in form oval.

GENUS **GALEOSCOPTES** CABANIS.

B. 254.   R. 12.   C. 16.   G. 8.   U. 704.

307. Galeoscoptes carolinensís (LINN.). Catbird. Summer resident; abundant in eastern Kansas. Arrive the last of April to the first of May. Begin laying about the middle of May. Nest usually on bushes at the edge of small thickets; composed of leaves, weeds, and strippings from grape-vines and stems of plants, and lined with rootlets. Eggs, from four to five; .94x.68; deep-bluish green; in form rather pointed.

GENUS **HARPORHYNCHUS** CABANIS.

SUBGENUS **METHRIOPTERUS** REICHENBACH.

B. 261, 261a.   R. 13.   C. 17.   G. 9.   U. 705.

308. Harporhynchus rufus (LINN.). Brown Thrasher. Summer resident; abundant. Arrive the first of April. Begin laying about the tenth of May. Nest in low bushes, vines and hedges — a coarse, bulky structure — base made usually of sticks, roots and stems of weeds; within this an inner nest of leaves and strippings from plants, lined with fine rootlets and horse hair. Eggs, four or five; 1.06x.80; ground color, white to light green, thickly dotted with reddish brown, sometimes yellowish brown; confluent around large end; in form rather elliptical.

SUBFAMILY **TROGLODYTINÆ**. WRENS.

GENUS **SALPINCTES** CABANIS.

B. 264.   R. 58.   C. 65.   G. 22.   U. 715.

309. Salpinctes obsoletus (SAY). Rock Wren. Summer resident in middle and western Kansas; common in suitable locations. Arrive in April. Begin laying the first of May. Nest in crevices in rocky ledges; loosely constructed of weeds, strippings of plants, grasses, bits of moss, wool, hair; in fact, any available substance, and often almost wholly of one kind. Eggs, four to nine; .74x.62; crystal white, sparingly specked with reddish brown, chiefly aggregating at and forming a wreath around large end; in form rather oval.

Genus THRYOTHORUS Vieillot.

Subgenus THRYOTHORUS.

B. 265.   R. 60.   C. 68.   G. 23.   U. 718.

310. Thryothorus ludovicianus (Lath.). Carolina Wren. Resident; abundant in eastern, rare in western Kansas. Begin laying early in April. Nest in crevices in old logs, rocks, and outbuildings; made of bits of twigs, grasses, and leaves, and lined with hair and a few feathers; quite bulky, usually filling the space, but when it is too high to fill, partially roof the nest over, entering a hole in the side. Eggs, five to seven; .74x.60; white, dotted pretty evenly and thickly over the surface with reddish brown, but sometimes thickest and forming a confluent band around the large end; in form rather oval.

Subgenus THRYOMANES Sclater.

B. 267.   R. 61.   C. 71.   G. 24.   U. 719.

311. Thryothorus Bewickii (Aud.). Bewick's Wren. Entered in first catalogue as "visitant and occasional resident in southern Kansas." Further examination proves the specimens I have captured and seen to be *variety bairdi;* but as I am not the only one that has reported the bird in the State, I will let it stand as entered, adding that in my opinion the birds do occasionally enter, and will be found in eastern Kansas. Nesting habits and eggs similar to Baird's Wren.

B. —.   R. 61*b*.   C. 72.   G. —.   U. 719*b*.

312. Thryothorus bewickii bairdi (Salv. & Godm.). Baird's Wren. Resident; not uncommon in southwestern Kansas. Nest in deserted woodpecker holes, hollow logs, or any nook it may fancy; composed of sticks, roots, straws and grasses, and lined with fur and a few downy feathers; quite bulky, generally filling the space, but in no case, I think, roofed over. Measurements of five eggs, taken at Corpus Christi, Texas, May 9th, 1882: .63x.50; .63x.50; .63x.50; .63x.49; .62x.49; white, speckled with light and dark shades of reddish brown, thickest around large end; in form oval.

Genus TROGLODYTES Vieillot.

Subgenus TROGLODYTES.

B. 271.   R. 63*a*.   C. 75.   G. 25, 26.   U. 721*a*.

313. Troglodytes aedon parkmanii (Aud.). Parkman's Wren. Summer resident; common. Arrive in April. Begin laying about the middle of May. Nest in holes in logs and stumps, and about dwelling houses in boxes, entering outhouses through crevices and knot-holes — in fact, most anywhere; I once found a nest in the skull of a buffalo; loosely constructed of sticks, weeds, etc., filling the cavities, leaving a small opening for entrance; within the rubbish they construct an inner nest, composed of finer material, lining the same with feathers, fur, and most any soft, warm substance. Eggs, seven to nine; .64x.49; ground color white, but so thickly dotted with specks of reddish brown and a few purple slate markings that the white is concealed; in form oval. In color and habits, a *fac simile* of the House Wren, *T. aëdon.* Our western specimens would naturally have a somewhat faded or bleached look when compared with eastern specimens, but I do not think that this alone—so slight and gradual a difference—should entitle it to rank as a subspecies; but I bow to the decision of the A. O. U. Committee, and drop, as a Kansas bird, *T. aëdon* from the list.

SUBGENUS **ANORTHURA** RENNIE.

B. 273.　R. 65.　C. 76.　G. 27.　U. 722.

314. Troglodytes hiemalis VIEILL. Winter Wren. Winter sojourner; rare. Leave the first of March.

GENUS **CISTOTHORUS** CABANIS.

SUBGENUS **CISTOTHORUS.**

B. 269.　R. 68.　C. 81.　G. 29.　U. 724.

315. Cistothorus stellaris (LICHT.). Short-billed Marsh Wren. Migratory; rare. Possibly breed in the State. Arrive in May.

SUBGENUS **TELMATODYTES** CABANIS.

B. 268.　R. 67, 67a.　C. 79, 80.　G. 28.　U. 725.

316. Cistothorus palustris (WILS.). Long-billed Marsh Wren. Summer resident; rare; in migration common. Arrive the last of April to first of May. Begin laying the last of May. Nest sometimes on a low bush, but generally in thick standing grass, on low, wet boggy marshes; made of leaves from the grasses, woven in and around the standing growing stalks; a globular nest, about five inches in diameter, with a small, round hole on the side for entrance, lined with feathers and a soft, cotton-like substance from plants. Eggs, five to nine; .63x.50; ground color ashy brown, but so thickly specked and blotched with deep chocolate brown that some specimens appear uniform; in form oval.

FAMILY **CERTHIIDÆ.** CREEPERS.

GENUS **CERTHIA** LINNÆUS.

B. 275.　R. 55.　C. 62.　G. 21.　U. 726.

317. Certhia familiaris americana (BONAP.). Brown Creeper. Winter sojourner; common. Leave in April.

FAMILY **PARIDÆ.** NUTHATCHES AND TITS.

SUBFAMILY **SITTINÆ.** NUTHATCHES.

GENUS **SITTA** LINNÆUS.

B. 277.　R. 51.　C. 57.　G. 19.　U. 727.

318. Sitta carolinensis LATH. White-breasted Nuthatch. Resident; common. Begin laying about the last of April. Nest in decayed hollows in trees; entrance a knot-hole or small opening; composed chiefly of rabbits' fur; in some cases a few fine grass leaves and feathers. Eggs, four to seven; .78x.58; rosy white, thickly specked, spotted and blotched with reddish brown, intermixed with purple; in form oval.

B. 279. R. 52. C. 59. G. 20. U. 728.

319. Sitta canadensis Linn. Red-breasted Nuthatch. Migratory; rare. Arrive in March to first of April.

Subfamily **PARIN/E.** Titmice.

Genus **PARUS** Linnæus.

Subgenus **LOPHOPHANES** Kaup.

B. 285. R. 36. C. 40. G. 16. U. 731.

320. Parus bicolor Linn. Tufted Titmouse. Resident; abundant in eastern Kansas. Begin laying about the middle of April. Nest in deserted woodpecker holes and natural cavities in trees; made of leaves and moss; lined with a fine, soft fibrous, cotton-like substance, and hairs from cattle. Eggs, five to eight; .75x.54; white, sprinkled with rusty red, thickest and somewhat running together around large end, with here and there a few blotches of lilac; in form oval.

Subgenus **PARUS** Linnæus.

B. 290. R. 41. C. 44. G. 17. U. 735.

321. Parus atricapillus Linn. Chickadee. Resident;. common. Begin laying early in April. Nest near the ground in holes made by themselves in decaying trees and stumps. Composed of bits of moss, interwoven with fur and fine hair, and occasionally a few feathers. Eggs, four to eight; .60x.47; white, specked with reddish brown and purple stains, generally thickest and forming a ring around large end; in form oval.

B. 289, 289a. R. 41. C. 45. G. 18. U. 735a.

322. Parus atricapillus septentrionalis (Harris). Long-tailed Chickadee. Resident; rare in eastern Kansas, but quite common in the western part of the State along the streams where skirted with trees and brush. Nesting habits and eggs (.63x.49) similar to *P. atricapillus.*

Family **SYLVIID/E.** Warblers, Kinglets, Gnatcatchers.

Subfamily **REGULIN/E.** Kinglets.

Genus **REGULUS** Cuvier.

B. 162. R. 33. C. 34. G. 15. U. 748.

323. Regulus satrapa Licht. Golden-crowned Kinglet. Winter sojourner; rare; in migration common.

B. 161. R. 30. C. 33. G. 14. U. 749.

324. Regulus calendula (Linn.). Ruby-crowned Kinglet. An occasional winter sojourner; in migration common.

Subfamily **POLIOPTILINÆ.** Gnatcatchers.

Genus **POLIOPTILA** Sclater.

B. 282.   R. 27.   C. 36.   G. 13.   U. 751.

325. Polioptila cærulea (Linn.). Blue-gray Gnatcatcher. Summer resident; rare; in migration common. Arrive in April. Begin laying the first of May. Nest in the branches of tree-tops, fifteen to fifty feet from the ground; usually saddled between and woven to upright twigs, a beautiful nest composed of stem-like stemlets, bits of leaves and feathers woven together with spiderwebs, lined with a soft downy substance from plants, and thickly dotted on the outside with lichens. Eggs, four or five; .56x.44; pale greenish white, spotted and blotched with reddish brown, lilac, and slate, running together around large end; in form rounded oval.

Family **TURDIDÆ.** Thrushes, Solitaires, Stonechats, Bluebirds, etc.

Subfamily **MYADESTINÆ.** Solitaires.

Genus **MYADESTES** Swainson.

B. 235.   R. 25.   C. 169.   G. 12.   U. 754.

326. Myadestes townsendii (Aud.). Townsend's Solitaire. An occasional fall and winter visitant in Western Kansas.

Subfamily **TURDINÆ.** Thrushes.

Genus **TURDUS** Linnæus.

Subgenus **HYLOCICHLA** Baird.

B. 148.   R. 1.   C. 6.   G. 1.   U. 755.

327. Turdus mustelinus Gmel. Wood Thrush. Summer resident; abundant in eastern Kansas. Arrive the last of April to first of May. Begin laying about the middle of May. Nest usually saddled onto a horizontal limb of a tree, six to ten feet from the ground, composed outside of loose stems of weeds and leaves, attached to a closely compact body of pulverized leaves, fibers and lint-like substances from plants, plastered together with saliva and tracings of mud, and lined with small fibrous roots. Eggs, three to five; 1.00x.75; deep blue with a greenish tinge; in form oval.

B. 151.   R. 2.   C. 7.   G. 2.   U. 756.

328. Turdus fuscescens Steph. Wilson's Thrush. Migratory; rare. Arrive the last of April to first of May. Probably breed in the State.

B. 54.   R. 3.   C. 12.   G. 3.   U. 557.

329. Turdus aliciæ Baird. Gray-cheeked Thrush. Migratory; rare. Arrive the last of April to the first of May.

B. 153. R. 4*a*. C. 13. G. 4. U. 758*a*.

330. **Turdus ustulatus swainsonii** (CAB.). Olive-backed Thrush. Migratory; common. Arrive the last of April to first of May.

B. 149. R. 5*b*. C. 10. G. 5. U. 759*b*.

331. **Turdus aonalaschkæ pallasii** (CAB.). Hermit Thrush. Migratory; rare. Arrive in April.

GENUS **MERULA** LEACH.

B. 155. R. 7. C. 1. G. 6. U. 761.

332. **Merula migratoria** (LINN.). American Robin. Resident; abundant in eastern Kansas, following up the settlements, and breeding throughout the State. Begin laying last of April. Nest in trees, hedges, outbuildings — in fact most anywhere off the ground — coarsely constructed of leaves, stems and grasses fastened together and plastered inside with mud, and lined with fine stems and rootlets. Eggs, four or five; 1.16x.80; greenish blue; in form oval.

B. —. R. 7*a*. C. 2. G. 7. U. 761*a*.

333. **Merula migratoria propinqua** RIDGW. Western Robin. A rare visitant in western Kansas. October 12, 1883, I killed two of the birds out of a flock of seven, at Wallace, Kansas.

GENUS **SIALIA** SWAINSON.

B. 158. R. 22. C. 27. G. 10. U. 766.

334. **Sialia sialis** (LINN.). Bluebird. An abundant resident in eastern Kansas, retiring in winter to the thickets along the streams. A rare summer resident in the western portion of the State, but increasing with its settlements. Begin laying about the last of April. Nest in bird-boxes, holes in trees and posts, loosely but rather smoothly constructed of fine grasses, with occasionally leaves, hairs, and a few feathers. Eggs, four or five; .83x.66; light blue; in form oval.

B. 160. R. 24. C. 29. G. 11. U. 768.

335. **Sialia arctica** (SWAINS.). Mountain Bluebird. Winter sojourner; rare in eastern Kansas.

ENGLISH SPARROW.

EUROPEAN HOUSE SPARROW—*Passer domesticus* (Linn.). LEACH.

In the Catalogue of 1883 I briefly expressed my views in regard to this bird as follows:

"The introduction of these hardy, aggressive little foreigners, with a view to their naturalization, has proved a decided success, and is therefore no longer a question of survival, but rather one as to whether the good they may do will overbalance the harm. In this discussion the birds will have their friends, but as a whole, those that know them best can but look forward with alarm at their rapid increase and spread. They cannot properly be classed with our native birds, but as they have come to stay, are entitled to a place or mention in our catalogues."

I am now able to lay before the reader the conclusions reached by the American Ornithologists' Union. At a meeting of the society, convened in the city of New York, September 26, 1883, a committee was appointed "to investigate the eligibility or ineligibility of the European House Sparrow." The following is the final report of the committee, as approved and adopted by the council of the American Ornithologists' Union, at Washington, April 21st, 1885:

Mr. President, and Members of the Union:—Your committee appointed to inquire into the eligibility of the European house sparrow (*Passer domesticus*) as a naturalized resident in this country, has the honor herewith to submit its report. After due consideration, your committee adopted the following form of circular letter, which was framed to elicit information from all quarters and from all interested persons:

The American Ornithologists' Union, an organization resembling the British association of similar name, and including in its active membership the most prominent ornithologists of the United States and Canada, purposes, among other objects already engaging its attention, to determine as nearly as possible the true status in America of the European house sparrow (*Passer domesticus*), commonly known as the English sparrow, in so far as the relations of this bird to mankind are concerned. The Union hopes to secure through the solicited testimony of others, as well as the personal observations of its members, the facts necessary to settle the question of the eligibility or ineligibility of this sparrow as a naturalized resident of this country. The question of the European house sparrow in America is regarded as one of great economic consequence, to be determined primarily by ascertaining whether this bird be, upon the whole, directly or indirectly injurious or beneficial to agriculture and horticulture. Its economic relations depend directly and mainly upon the nature of its food; indirectly upon the effect, if any, which its presence may have on useful native birds and beneficial insects. The accompanying formula of questions is respectfully submitted to the attention of those who may be able and willing to record statements of positive facts and value derived from their own experience. Concise and unquestionable answers returned to the undersigned on inclosed blank, or otherwise, or communicated to any member of the committee, will be appreciated, and prove of high value among the data upon which it is hoped that this vexed question may be set at rest. The evidence thus obtained will be carefully considered by the committee in preparing its report to the Council of the Union, and a digest of the same, with recommendations, if any, will be submitted by the Council to the mature judgment of the Union at its next annual meeting. The following-named active members of the Union were, at the first congress, appointed a committee to investigate and report upon this subject: Dr. J. B.

Holder, of New York, chairman; Mr. Eugene P. Bicknell, of New York; Mr. H. A. Purdie, of Boston, Mass.; Mr. Nathan Clifford Brown, of Portland, Me.; Mr. Montague Chamberlain, of St. John, New Brunswick; the committee having the power of increasing its membership at its discretion.

DR. J. B. HOLDER, *Chairman.*

AMERICAN MUSEUM OF NATURAL HISTORY, Central Park, New York City, Feb. 2, 1884.

DATA CONCERNING THE EUROPEAN HOUSE SPARROW, FROM..................

1. Is the European house sparrow (*Passer domesticus*) known in your neighborhood, and if so, about when did it appear? 2. Is your neighborhood city, suburbs, or country? 3. Is this sparrow abundant? 4. Is it increasing in numbers? 5. How many broods and young, yearly, to a pair? 6. Is this sparrow protected by law? 7. Is it artificially fed and housed? 8. Does it molest, drive away or diminish the numbers of native birds? 9. If so, what species? 10. Does this sparrow injure shade, fruit or ornamental trees? 11. Does it attack or injure garden fruits and vegetables? 12. Does it injure grain crops? 13. Is it an insect-eater or a seed-eater? 14. What insects, if any, are chiefly eaten by this sparrow? 15. What is the principal food it carries to its young? 16. What insects, if any, are carried by it to its young? 17. Does the food of the old bird vary with the seasons, and if so, in what way? 18. Does the food of its young vary, and if so, how? 19. If any insects are eaten, are they beneficial or injurious species? 20. Does the sparrow eat the larvæ of the vaporer moth (*Orzvia leucostigma*)? 21. Does it eat ichneumon flies? 22. Do you determine the nature of this bird's food and that furnished by it to its young, by inference, direct observation, or dissection? 23. Have any injurious insects been exterminated or materially lessened in numbers by this sparrow? 24. Have any injurious insects increased in numbers, or appeared where unknown before, in consequence of the destruction of other insects by this sparrow? 25. Have these sparrows in your neighborhood been destroyed systematically or otherwise, and if so, by what means? 26. What bounty, if any, has been offered for their destruction? 27. What is the general sentiment or balance of public opinion respecting the European house sparrow in your locality? 28. On the whole, in your judgment, is this sparrow an eligible or ineligible species in this country?

In order to secure a thorough presentation of the subject to those most likely to respond satisfactorily, each member of the committee assumed the duties of correspondence in his own section of the country, as well as in certain allotted sections of the entire United States and Canada. Copies of the letter were sent to the agricultural papers, to the various journals having columns devoted to zoölogical and rural matters, and to the press at large. The greater part, however, was directed to individuals believed to possess facts pertinent to the subject. About one thousand copies were thus sent out.

A large proportion of the answers received are of one import, written by persons having no definite data to communicate, but who, having experienced annoyance from the bird's uncleanliness and unmusical notes, desire to see it exterminated. Under this head belong the numerous petitions which have reached us from several quarters, notably from Philadelphia. The subject is regarded sufficiently important by the inhabitants of that city to warrant the issue of printed forms, which, with long lists of subscribers, have been submitted to the consideration of your committee. The paucity of replies to many of our questions renders it impossible to report upon them decisively. Fortunately, however, others are very fully answered.

Returns to the first question give some data of interest in relation to the time of the sparrow's first introduction into this country. The earliest date of importation known to us is 1858, when Mr. Thomas A. Deblois liberated a few individuals at Portland, Me. These disappeared shortly afterward, and were not successfully replaced until 1875. In 1858 sparrows were liberated at Peacedale, R. I., by Mr. Joseph Peace Hazard. They were first introduced into Central Park, New York city, according to Mr. Conklin, the Superintendent of the Menagerie, in the year 1864. In 1860 Mr. Eugene Shiefflin turned loose twelve birds in Madison Square, New York city. In 1868 the species was first introduced into Boston Common. In 1869 a number were given the liberty of the parks of Philadelphia. Somewhat later a successful attempt was made to establish a colony near Great Salt Lake, Utah, and about the same time the birds became resident at Indianapolis, Ind.

In a period of about ten years, the sparrows reached nearly all the large towns and cities of New England and the Middle States, and many of those of the Western States, without artificial assistance. It also made its appearance in suburban towns, and even country villages. From the Southern States, and the Western States beyond the Mississippi river, we have received but few returns, and most of these state that the sparrow has not been observed. In Canada it has become generally distributed over the southern sections of Quebec and Ontario, (it is abundant in the city of Quebec,) and in 1884 several flocks invaded New Brunswick.

Few observers have definitely determined the number of broods hatched yearly by this bird, and the number of young to the brood. We have, however, returns from several ornithologists. The maximum given by Mr. H. B. Bailey, of Orange, New Jersey—six broods in one season, with from four to five young in a brood — probably indicates the extent of the bird's fertility in this climate. The usual number of broods in the latitude of New York and southward appears to be four. In more northern districts, three broods yearly would probably be near the average.

There is an overwhelming mass of testimony to the effect that the sparrow molests and drives away certain of our most valued species of native birds. Many statements have been received, giving accounts of conflicts provoked by the sparrow, in which it was cruelly victorious. It is affirmed that from some localities native species have been completely banished by the attacks or by the mere presence of the foreigner. We have also evidence of an opposite character, declaring the sparrow's peaceable disposition, and its association upon amicable terms with other species of birds.

Most of our correspondents state that they have never known the sparrow to commit depredations upon crops, but well-authenticated instances are furnished showing its ability and disposition to accomplish great destruction to grain. Mr. Stewart, of Hackensack, New Jersey, relates the destruction of a wide margin of wheat in the field. Hon. G. A. Bicknell, of New Albany, Indiana, says: "When the grain ripens, the sparrows leave the city and attack the wheat fields in the suburbs. I have seen hundreds of them at once in my fields, and they got about half the crop." Mr. T. G. Gentry, in his exhaustive work on the sparrow, gives similar instances. That the bird feeds upon fruits, is amply attested.

Our thirteenth question calls for information as to the sparrow's preference for food. Is it an insect-eater or a seed-eater? Every reply to this question, which is based upon dissection, agrees in attributing to the bird a diet almost wholly vegetable. The statement of some observers, that it devours canker worms and a variety of insects, is unaccompanied by reports of examinations of the stomach.

The question as to the food of nestling sparrows elicited pretty uniform testimony, animal matter in some form being said to constitute the bulk. Dissections by a competent person, however, show "barely a trace of insect or animal food, but in lieu, fine gravel and vegetable fiber."

Responses to questions seventeen to twenty-one, inclusive, are too meager to be of value.

It is claimed by several of our correspondents that the measuring worm, so abundant at the time of the sparrow's introduction into this country, was well-nigh exterminated by the bird, so that for a considerable period it was unobserved. Since it is a well-known fact that the worm occurs in very variable numbers in different seasons, credit for its comparative extermination in this case can hardly be given to the sparrow upon the doubtful evidence before us.

The experiment has recently been tried in Philadelphia and elsewhere of substituting sparrows for pigeons in trap-shooting, but, of course, without seriously diminishing their numbers. In other localities the birds have been poisoned or otherwise gotten rid of to some extent by indignant citizens in defiance of laws.

The balance of public opinion is strongly adverse to the sparrows. Our returns, however, show protective laws (usually the same statute which provides for the security of other small birds) in Maine, New Hampshire, Vermont, Rhode Island, New York, New Jersey, Ohio, Michigan, the District of Columbia, and Canada. The Massachusetts law has lately been repealed, and specially exempts the English sparrow from protection.

So much for the evidence. We have learned the capacity and disposition of this bird to injure grain and fruits, and that when gathered in large numbers it threatens very seriously the interests of the farmer and horticulturist. Although testimony of a certain kind indicates that its young are fed with insects, actual dissection shows that vegetable substances are mainly employed. The adult birds feed almost exclusively upon seeds and grains. They drive away from their accustomed haunts, either directly or indirectly, many of our native insectivorous species. It may be added that they have proved in recent years so destructive of crops in other countries, as to render it necessary to enact laws looking to their extermination. In view of these facts, your committee believes that the European sparrow (*Passer domesticus*) is an ineligible species in this country, and that it was a mistaken policy to introduce the bird. And we would respectfully recommend:

1. That sheltering or otherwise fostering the sparrow by the public be discouraged, and that its introduction artificially in new localities and its sale for such purposes be forbidden by law.

2. That all existing laws protecting the sparrow be repealed, and that bounties be offered for its destruction.   (Signed)   J. B. HOLDER, *Chairman.*
EUGENE P. BICKNELL.
H. A. PURDIE.
NATHAN CLIFFORD BROWN.
MONTAGUE CHAMBERLAIN.

PROTECT THE BIRDS!

As this work will fall largely into the hands of those interested in birds and bird-life, I think much good may be done toward furthering the efforts of the "American Ornithologists' Union Committee on the Protection of North American Birds," by publishing the following, clipped from "Bulletin No. 1," as explanatory of their objects. And in doing so, I most earnestly appeal to parents, teachers, ministers, and the leaders of the various societies, especially the ladies' societies, to awaken to the fact that unless the birds, so valuable as a check upon insect-life, and whose presence and song gladden the heart, are protected, they will soon become exterminated, like our "buffalo," that so short a time ago were in almost countless numbers upon our prairies, slaughtered, as our native birds are now, for the sport and their skins. Just in the ratio as the birds decrease, will the price set upon their heads for their plumage increase, unless the craze for feathers as ornaments can be checked. It is therefore time for action, and I feel confident our appeal will not be in vain.

AMERICAN ORNITHOLOGISTS' UNION COMMITTEE ON BIRD PROTECTION.

The American Ornithologists' Union Committee was recently organized in New York city, with the following membership: Mr. George B. Sennett, chairman; Mr. Eugene P. Bicknell, secretary; Mr. Willian Dutcher, treasurer; Mr. J. A. Allen, Dr. J. B. Holder, Dr. George Bird Grinnell, and Mr. L. S. Foster, all of New York city; Mr. William Brewster, Cambridge, Mass.; Mr. Montague Chamberlain, St. John, N. B.; Col. N. S. Goss, Topeka, Kan.

The committee is desirous of collecting facts and statistics bearing upon the subject of the destruction of our birds, and will welcome information from any source.

It also extends the promise of its hearty coöperation to all persons or societies who may be interested in the protection of birds.

The headquarters of the committee are at the American Museum of Natural History, Central Park, New York city, where the officers or any of the members may be addressed.

THE AUDUBON SOCIETY.

In order to give an opportunity for definite and systematic effort by all those who believe that our birds ought to be protected, the *Forest and Stream* has recently founded the Audubon Society. Membership in this society is to be free to every one who is willing to assist in forwarding any one of the three objects for which it is established. These objects are to prevent so far as possible (1) the killing of any wild bird not used for food; (2) the destruction of the nests or eggs of any wild bird; and (3) the wearing of feathers as ornaments. *The work to be done by the Audubon Society is auxiliary to that which is being done by the American Ornithologists' Union Committee,* and will consist largely of matters of detail, to which this committee could not attend. The management of the society for the present will be in the hands of a member of this committee. Branches of this association will be established all over the country. The work of the *Forest and Stream* is only preliminary. As soon as the society shall have attained a respectable membership, and be on a firm footing, it will be turned over to its members for final organization. In order that this may take place as speedily as possible, it is hoped that all interested in bird protection will send in for membership their own names, as well as those of any others whom they think likely to assist. To all such, free circulars containing information will be sent for distribution. Names should be sent without delay to *Forest and Stream,* 40 Park Row, New York, N. Y.

BIRD LAWS.

Most of the States and Territories have on their statute books laws for the protection of game and fish, regulating the season of hunting and fishing, and providing penalties for the taking of game or fish during certain portions of each year, or, in particular cases, for a series of years. These laws are intended, in most cases, to give protection to "useful" birds, in addition to the game birds, and their nests and eggs, at all seasons. In general, these laws are crude and unsatisfactory so far as they relate to supposed useful birds, and also in relation to many others which are either protected merely during certain months, or not at all, as is the case with many of the marsh and shore-inhabiting species, such as the herons, terns, gulls, etc. Most of the laws exclude from protection all hawks and owls, crows, jays, and blackbirds, and, in some cases, robins and other kinds of song birds, woodpeckers, etc. A few of the laws make provision for collecting birds and their eggs for scientific purposes, often in a lax way, but occasionally, as in Maine, with considerable stringency; while the new bird law of New Jersey prohibits the destruction of song birds, their nests or eggs, for any purpose whatever. Defective as the present laws now generally are, they would, if thoroughly enforced, prevent the disgraceful slaughter now so general and untrammeled by any legal interference. As already so many times reiterated in this series of papers, the fault is not so much lack of laws, or inadequate legislation, as the absence of nearly all effort to interpose any obstacles, legal or otherwise, in the way of free slaughter. So apathetic is the public in all that relates to bird protection, that prosecution under the bird-protection statutes requires, on the part of the prosecutor, a considerable amount of moral courage to face the frown of public opinion, the malignment of motive and the enmities such prosecution is sure to engender.

None of the bird laws are above improvement, even in so far as they relate to the protection of game birds; but, in respect to the non-game birds, nearly all require more or less change. If possible, it would be well to have uniform laws throughout all the States and Territories, varying only in respect to the time of the close season, and such other points as difference of season, kind of game to be especially protected, etc., according to local conditions. At present certain birds are protected in some States which are outlawed in others, or are treated as game birds in some and not so treated in others.

Birds, as regards legislation, may well be divided into two classes — game birds, and birds which are not such; and the laws relating to each class should be separate and distinct. The game birds should be left to the care of sportsmen and game protective associations, since self-interest on the part of the more intelligent sportsmen will dictate more or less wise legislation for the preservation of the birds on which their sport depends. But in respect to game birds, public opinion should be so far enlightened as to secure the enforcement of proper legislative enactments, which is notoriously not the case at present. All other birds should be left to the care of bird-lovers and humanitarians, who should see that proper laws for their preservation are not only enacted, but duly enforced. As already shown in preceding pages of this supplement, those who know best, from having scientifically investigated the subject, are convinced that none of our native birds should be outlawed as unqualifiedly, or even to any serious degree, injurious. A few exceptions might be made, were it practicable, but in the general ignorance of legislators and of the public generally, or their inability to make proper distinction through inability to recognize by proper names one kind of hawk, for instance, from another, the safe way is to attempt no such discrimination in legislation. The slight harm resulting from protecting half a dozen species more or less harmful would be more than offset by the indiscriminate destruction which would necessarily result from such a loophole.

The reason for keeping legislation respecting game birds distinct from that relating to the other species is mainly to avoid conflict of interests respecting such legislation, which is more or less sure to follow in any attempt at combined legislation respecting all birds in one act. Sportsmen's clubs and game protective associations in attempting to provide proper game laws often find strong opponents in the game-dealers and market-gunners, who often succeed in defeating judicious legislation. If all birds are treated under the same act, attempts to improve the portions of such acts as relate to useful birds are often prevented through opposition to certain clauses of the game sections obnoxious to pot-hunters and game-dealers, as has recently been the case with attempted judicious amendments to the bird laws in the State of Massachusetts.

There should also be some provision for collecting birds, their nests and eggs, for scientific purposes, in behalf of our natural history museums, and of scientific progress in ornithology. As already shown in these articles, the birds destroyed in the interest of science, notwithstanding the outcry to the contrary from certain sources, are relatively few in comparison to the number destroyed for millinery and other mercenary purposes — so small as not to materially affect the decrease of any species. But such license, unless rigidly guarded, is liable to abuse, and should be hedged about with every practicable safeguard. The number of such licenses issued in any State should be very small; they should be granted with strictest regard to the fitness of the recipient to be allowed such a favor; and their abuse or misuse made a misdemeanor, subject to severe penalties. Obviously, the power to grant them should, so far as possible, be vested in persons having some knowledge of ornithology, or who are able to recognize the difference between collecting birds for scientific purposes and as "curiosities," or for traffic other than strictly in the interest of science. It should be further understood that these licenses grant no immunity from the ordinary laws of trespass, or laws against the use of firearms at improper times or places, or in violation of any of the provisions of game protective acts. The system of issuing such licenses has needlessly been brought into disrepute through the gross ignorance and apathy of the general public as to their real purpose and limitations. For most of the abuses of the system there is already abundant remedy. Any person holding such a license, who uses it as a shield against prosecution for illegal or indiscriminate slaughter of birds for any and all purposes, is successful only to such extent as the ignorance or apathy of the community among which his misdeeds are committed happen to give him immunity. The fault is not in reality chargeable to the law, or the system permitting the granting of certificates for scientific collecting. In this matter, as in all else relating to bird destruction, all that is necessary to prevent abuses is a proper comprehension of the laws relating to the subject, and a public sentiment not only favorable to their enforcement, but watchful against any infringement of their provisions.

With a desire to bring about more intelligent, uniform and desirable legislation for the protection everywhere, and at all times, of all birds not properly to be regarded as game birds, the American Ornithologists' Union Committee on bird protection have had under careful consideration a draught of a bird law, drawn with special reference to its fitness for general adoption throughout the United States and the British Provinces, and with regard to just what birds should be so protected. It is intended as a guide or model, which may serve as a basis for legislation. From its pertinence in the present connection, it is given below in full. Possibly some additional provisions may still be desirable, relating especially to the designation of certain officers to secure its strict observance, the amount of the fine, and whether or not a part of the fine should go to the complainant; features, however, that doubtless may be safely left to legislative discretion.

AN ACT FOR THE PROTECTION OF BIRDS AND THEIR NESTS AND EGGS.

SECTION 1. Any person who shall, within the State of —————, kill any wild bird other than a game bird, or purchase, offer, or expose for sale any such wild bird, after it has been killed, shall for each offense be subject to a fine of five dollars, or imprisonment for ten days, or both, at the discretion of the court. For the purposes of this act the following only shall be considered game birds: The Anatidæ, commonly known as swans, geese, brant, and river and sea ducks; the Rallidæ, commonly known as rails, coots, mud-hens, and gallinules; the Limicolæ, commonly known as shore-birds, plovers, surf-birds, snipe, woodcock, sandpipers, tatlers, and curlews; the Gallinæ, commonly known as wild turkeys, grouse, prairie chickens, pheasants, partridges, and quails.

SEC. 2. Any person who shall, within the State of —————, take or needlessly destroy the nest or the eggs of any wild bird, shall be subject for each offense to a fine of five dollars, or imprisonment for ten days, or both, at the discretion of the court.

SEC. 3. Sections 1 and 2 of this act shall not apply to any person holding a certificate giving the right to take birds, and their nests and eggs, for scientific purposes, as provided for in section 4 of this act.

SEC. 4. Certificates may be granted by [here follow the names of the persons, if any, duly authorized by this act to grant such certificates], or by any incorporated society of natural history in the State, through such persons or officers as said society may designate, to any properly accredited person of the age of eighteen years or upward, permitting the holder thereof to collect birds, their nests or eggs, for strictly scientific purposes only. In order to obtain such certificate, the applicant for the same must present to the person or persons having the power to grant said certificate, written testimonials from two well-known scientific men, certifying to the good character and fitness of said applicant to be intrusted with such privilege; must pay to said persons and officers one dollar to defray the necessary expenses attending the granting of such certificates; and must file with said persons or officers a properly executed bond, in the sum of two hundred dollars, signed by two responsible citizens of the State as sureties. This bond shall be forfeited to the State, and the certificate become void, upon proof that the holder of such a certificate has killed any bird, or taken the nest or eggs of any bird, for other than the purposes named in sections 3 and 4 of this act, and shall be further subject for each such offense to the penalties provided therefor in sections 1 and 2 of this act.

SEC. 5. The certificates authorized by this act shall be in force for one year only from the date of their issue, and shall not be transferable.

SEC. 6. The English or European house sparrow (*Passer domesticus*) is not included among the birds protected by this act.

SEC. 7. All acts or parts of acts heretofore passed, inconsistent with or contrary to the provisions of this act, are hereby repealed.

SEC. 8. This act shall take effect upon its passage.

AN APPEAL TO THE WOMEN OF THE COUNTRY ON BEHALF OF THE BIRDS.

The relation of the women of the country to the present lamentable destruction of bird-life has been several times alluded to in the foregoing pages; but the matter is so important, it demands more formal notice in the present connection. The destruction of millions of birds annually results from the present fashion of wearing birds on hats and bonnets. The women who wear them, and give countenance to the fashion, have doubtless done so thoughtlessly, as regards the serious destruction of bird-life thereby entailed, and without any appreciation of its extent or its results, considered from a practical standpoint. Until recently, very rarely has attention been called to the matter, or the facts in the case been adequately set forth. They have therefore sinned for the most part unwittingly, and are thus not seriously chargeable with blame. But the case is now different, and ignorance can no longer be urged in palliation of a barbarous fashion. Obviously it is only necessary to call the attention of intelligent women to the subject, as now presented, to enlist their sympathies and their efforts in suppression of the milliner's traffic in bird skins. As a recent writer (Mr. E. P. Bicknell, Secretary of the American Ornithologists' Union Committee on Bird Protection) in the *Evening Post* of this city, has not only forcibly appealed to the women in behalf of the birds, but suggested to them certain desirable lines of action, this brief reference to the subject may well be concluded with a few pertinent extracts from the article in question.

"So long as demand continues, the supply will come. Law of itself can be of little, perhaps of no ultimate, avail. It may give check; but this tide of destruction it is powerless to stay. The demand will be met; the offenders will find it worth while to dare the law. One thing only will stop this cruelty—the disapprobation of fashion. It is our women who hold this great power. Let our women say the word, and hundreds of thousands of bird-lives every year will be preserved. And until woman does use her influence, it is vain to hope that this nameless sacrifice will cease until it has worked out its own end, and the birds are gone. . . . It is earnestly hoped that the ladies of this city can be led to see this matter in its true light, and to take some pronounced stand in behalf of the birds, and against the prevailing fashions.

" It is known that even now birds are not worn by some, on grounds of humanity. Yet little is to be expected from individuals challenging the fashion—concert of action is needed. The sentiment of humanity once widely aroused, the birds are safe. Surely those who unthinkingly have been the sustaining cause of a great cruelty will not refuse their influence in abating it, now that they are awakened to the truth. Already word comes from London that women are taking up the work there. Can we do less? It needs only united action, sustained by resolution and sincerity of purpose, to crush a painful wrong—truly a barbarism—and to achieve a humane work so far-reaching in its effects as to outsweep the span of our own generation, and promise a blessing to those who will come after."

There are already in England, it may be added, two societies organized expressly in aid of the preservation of birds "in Great Britain and all other parts of the world." The Selborne society, originated by George Arthur Musgrave, of London, appeals to Englishwomen "to forswear the present fashion of wearing foreign or English bird skins. Our countrywomen are asked to inaugurate a return to a mode which, though half forgotten now, is assuredly more becoming to the wearer than trophies of robins and sandpipers." Lady Mount Temple is not only a member of the plumage section of the Selborne society, but has written a vigorous protest against the fashion of wearing dead birds on dresses, bonnets, and hats. The section is under the patronage of her Royal Highness the Princess Christian of Schleswig-Holstein, and numbers among its membership twenty ladies of title, and also Lord Tennyson, Robert Browning, Sir Frederick Leighton, and Rev. F. O. Morris.

INDEX.